THE *Albatross* AND THE *Fish*

MILDRED WYATT-WOLD SERIES IN ORNITHOLOGY

The ALBATROSS and the FISH

Linked Lives in the Open Seas

ROBIN W. DOUGHTY *and* VIRGINIA CARMICHAEL

Foreword by H.R.H. PRINCE OF WALES Introduction by JOHN CROXALL

UNIVERSITY OF TEXAS PRESS AUSTIN

Requests for permission to reproduce material
from this work should be sent to:
Permissions
University of Texas Press
P.O. Box 7819
Austin, TX 78713–7819
www.utexas.edu/utpress/about/bpermission.html

The paper used in this book meets the minimum requirements
of ANSI/NISO Z39.48–1992 (R1997) (Permanence of Paper). ∞

LIBRARY OF CONGRESS CATALOGING-IN-PUBLICATION DATA

Doughty, Robin W.
The albatross and the fish : linked lives in the open seas / Robin W. Doughty and
Virginia Carmichael ; foreword by H.R.H. Prince of Wales ;
introduction by John Croxall.
p. cm. — (Mildred Wyatt-Wold series in ornithology)
Includes bibliographical references and index.
ISBN 978-0-292-72682-6 (cloth : alk. paper) — ISBN 978-0-292-73763-1 (e-book)
1. Albatrosses—Conservation. 2. Albatrosses—Effect of chemicals on.
3. Albatrosses—Effect of pollution on. 4. Sea birds—Ecology.
5. Fishes—Conservation. I. Carmichael, Virginia. II. Title.
QL696.P63D68 2012
333.95842—dc23
2011021620

FRONTISPIECE: Angled Black-browed Albatross looks like a tipping
point: Can we save the albatross? Courtesy Greg Lasley.

I suspect that no other bird in the world will ever command such a profound image of avian grandeur and presence.

GRAHAM COLLIER,
ANTARCTIC ODYSSEY

Contents

THE ALBATROSS AND THE FISH

The continuing decline of the albatross is a parable of our times, and perhaps of our future. Decades after the direct threats to the remaining species appeared at last to be under control, the albatross family has become the most severely threatened bird group in the world.

The years of devastation caused by the ravages of whaling and sealing crews on nesting colonies were brought to an end in the 1980s. Yet instead of seeing the long-awaited increases in these fragile and slow-breeding populations, scientists began to note numbers declining. As these widely separated experts began piecing together the puzzle, they discovered that the new, highly efficient longline fishing technology was also highly efficient at hooking albatrosses. Tens of thousands of birds were being killed each year from longline fishing alone, not including deaths from trawl-fishing and, most insidiously, from illegal fishing operations around the globe. These tragic and unnecessary losses were causing most populations to decline rapidly, in many cases at rates that were so fast as to threaten the survival of the species.

It is now abundantly clear that the fate of the albatross family is linked to that of high-value table fish like bluefin tuna and Chilean sea bass, whose populations have also been dangerously reduced by overharvesting. The fate of such endangered fish is in turn tied to the way that humanity manages, or fails to manage, the natural abundance of the open seas.

This remarkable book tells an important story, revealing both the complexity of the issues and the urgency of the challenges to the marine environment. In describing the unlikely formation of an international and global environmental community to save the albatross, the authors have revealed essential truths about the relationships between seabirds, fish and ourselves. They make a powerful case for taking an ecosystem approach to human activity in such an interrelated world, and they point clearly to the roles each one of us will need to play if we are to make that a reality.

The fate of the oceans currently remains in the balance because of our own desires and habits as consumers; because of the legal and illegal drive for profit without concern for sustainability, and the devastating pollution of our oceans worldwide. In this urgent global situation, the albatross serves as an astonishingly beautiful and iconic reminder of what we are placing at risk. It would be an utterly shameful indictment of so-called civilized society if we could not find a way to live our lives in a way that would allow these magnificent birds to share the same planet with us. It would also call into question our ability to ensure our own survival on a planet of finite natural resources.

Acknowledgments

WE ARE EXTREMELY GRATEFUL to the many people who have generously offered time, interest, and expertise about the albatross and who clarified issues and thrilled us with their energy and commitment to these magnificent wild and mysterious birds. Nongovernmental organizations, such as BirdLife International, the Royal Society for the Protection of Birds (RSPB), National Audubon, Birds Australia, Aves Uruguay, Forest and Bird, Sociedad Española de Ornitología, and others, including their staffs and researchers, have supplied data and information from various perspectives about the albatross. Government personnel in Argentina, Australia, Canada, Chile, France, New Zealand, South Africa, Spain, the United Kingdom, and the United States and staff members and researchers of Regional Fisheries bodies, notably those of the Convention for the Conservation of Antarctic Marine Living Resources (CCAMLR), as well as those of the Agreement for the Conservation of Albatrosses and Petrels (ACAP), have supplied information about marine resources, fisheries statutes, and seabird mitigation techniques and have responded patiently to our queries. We appreciate their kindness. It is difficult to name all the people who have assisted us in this bold, indeed foolhardy, endeavor. John Croxall, lately of the British Antarctic Survey and now with BirdLife International, deserves our special gratitude for visiting with and guiding us, and for perusing various iterations of the manuscript. Cleo Small and Brian Sullivan have been most kind in discussing aspects of international fisheries developments. In Australia, we thank Rosemary Gales, Mike Double, Peter Milburn, Lindsey Smith, and others who supplied data about the birds. Barry Baker proved most helpful in briefing Robin, introducing him to a number of experts, and helping him attend the inaugural ACAP Conference of Parties in Hobart, Tasmania. Christopher Robertson, Neville Peat, and Lyndon Perriman have kindly supplied information about the albatrosses and researchers in New Zealand; Henri Weimerskirch and Pierre Jouventin have done so for

Indian Ocean islands. We are most grateful to the many persons who gave their encouragement, guidance, and expertise. They are: Sebastián Álvarez, Harry Battam, Steve Campbell, Carlos Carboneras, Agustín Carriquiry, John Cooper, Robert Crawford, Frances Cushing, Aidan Davison, Richard Dawson, Euan Dunn, Liza Fallon, Rob Fergus, Olivia Gable, Ben Galbraith, Nerina Gepp, Luke Gladd, Robert Hall, Hiroshi Hasegawa, Marcus Haward, Ian Hay, Neil Hermes, Glen Hurry, Douglas Hykle, Jamie Kirkpatrick, Shirley Kraal, Lorne Kriwoken, Greg Lasley, Rebecca Lewison, Isobel Lionaz, Brian Macdonald, Jeff Mangel, Ian Manners, Sandra McGrouther, James McKee, Karina McLachlan, Andrew McNee, Jonathan Meiberg, Ed Melvin, Peter Milburn, Denzil Miller, Janice Molloy, Ken Morgan, John O'Sullivan, Barbara Parmenter, Chris Perrins, Samantha Petersen, Lars Pomara, Kim Rivera, Christopher Robertson, Graham Robertson, Michael Short, Moulay Anwar Sounny-Slitine, Adrian Stagi, Elaine Stratford, Mark Tasker, Lance Tickell, Bill Wakefield, Eric Woehler, and Robin Woods.

The University of Texas at Austin, especially the Edward A. Clark Center for Australian and New Zealand Studies, through the gracious assistance of John Higley, has provided funds for research and travel. Faculty and staff in the Department of Geography and the Environment have been constructive critics as well as conscientious supporters of this project. The Australian National University and University of Tasmania at Hobart supplied logistical support as well as being excellent venues for substantive discussions and access to reference and archival materials. Edward Grey Institute's Alexander Library in the Department of Zoology at the University of Oxford offers an inexhaustible fund of accessible publications about birds, including albatrosses. We appreciate editor Craig Colten's inclusion of these lovely birds in a special issue of the *Geographical Review* titled "Avian Geography." The University of Texas Press editorial staff have been most accommodating to our various idiosyncrasies. We thank all of you.

Introduction

JOHN CROXALL

ALBATROSSES: despite the remoteness of their breeding sites and the formidable isolation of an oceanic lifestyle, their survival in the several centuries since we discovered them has been at increasing risk from the activities of mankind. Only now, at what could be the last opportunity for many albatross populations, are concerted efforts under way to redress the situation.

Three main interlinked philosophical and practical drivers have conspired to bring about the demise of albatrosses. First is the pervasive view in many (but not all) societies and cultures that wildlife on this planet exists only to serve human need and convenience. Thus, for breeding sites accessible to humans, this led to direct exploitation of adults, chicks, and eggs by the thousands for human consumption or adornment. At sea, such attitudes still contribute to the perception that killing albatrosses (and many other marine creatures) as by-products of fishing activity is acceptable and inconsequential. Another disastrous by-product of humankind's early interactions with albatross breeding sites was the introduction of alien invasive species. Although albatrosses are, in general, less vulnerable to such interactions than smaller birds and animals, they have suffered greatly from direct depredation by cats, rodents, pigs, etc., and indirectly from changes to breeding habitat wrought by other species such as goats and rabbits.

Second, human exploitation of resources, especially marine ones, being driven by maximizing immediate profit or return, has led to rapid and irreversible overfishing. This is part of the "tragedy of the commons," whereby foregoing today's short-term profit for tomorrow's long-term sustainable harvesting is seldom favored in competitive societies where supply is, or is perceived to be, outstripped by demand. This is particularly the case in systems where either exploitation is unregulated or, if nominally regulated, such rules are unenforced or unenforceable. Nowhere is this more apparent than on the high seas.

For generations now, this has been the last frontier, one where piracy flourishes, where any regulated practitioners have to compete (and therefore often collaborate) with exploiters unfettered by regulation, where ignoring and evading any rules is condoned by many governments and by the self-appointed management authorities alike, and where the rule (or rather the enforcement) of international law is so weak as to be largely ineffectual.

The third driver is the pressures on natural resources generated by the exponential growth of human population. This is compounded by the increasing disparity between the affluent and the poor and the demand for food resources to sustain the latter in the face of the increasingly powerful monopolies that, at sea, effectively control all but artisanal fishing practices. The imperative of poverty alleviation is widely championed but seldom linked to the need for population regulation, without which increasing poverty and resource demand are inevitable consequences.

Taken together, these attitudes, practices, and priorities combine to ensure the continuing overexploitation of the marine resources of our oceans, with little or no concern for the fate of the species and systems that naturally depend on these. In addition, one aspect that consistently compounds the plight of species such as albatrosses is the lack of appreciation of just how vulnerable many species of marine top predators actually are. Most albatross species breed on just one or a few islands, have global populations on the order of ten thousand pairs or fewer, and have some of the lowest reproductive rates among birds (they start breeding at ten years of age and often do not breed every year thereafter; they are successful on only about one-half of breeding occasions). Thus they are extremely sensitive to additional sources of mortality at any stage of their life cycle but especially when adult.

This extraordinary book provides detailed insights into all aspects of the threats confronting albatrosses, both historically, currently, and in prospect. In particular, however, it focuses on the current struggle to address the impact of bycatch in commercial fisheries, especially those for tuna and on the high seas. Thus, in the 1980s, just at a time when efforts were being made to preserve and protect the island habitats where most albatrosses breed, a devastating new threat emerged in the ocean habitats where they feed. Perhaps inevitably, in the search for new target species and more effective ways of fishing, but accelerated by the ban on driftnet fishing on the high seas, longline fishing began to cast a shadow over the life of albatrosses. Longline fishing rapidly developed into a preeminent fishing technique, whether targeting high-value fish in surface waters or fishing near the ocean floor at ever-increasing depths and in hitherto pristine environments, especially at seamounts. If this was rapidly a disaster for the survival of many of the target fish species, it was little less so for the marine mammals, turtles, and seabirds—and the nontarget fish species (especially sharks)—caught incidentally. Once the magnitude of the problem became appreciated (which took a decade or more, not least because of the low absolute rates of bycatch—until these were scaled up to take account of the huge fishing effort involved), an increasingly complex operation was started, designed to mitigate, if not eliminate, the bycatch of nontarget species, including and especially albatrosses.

As this book describes, the theaters of engagement were diverse and the actors multiple at every level from grassroots interaction with fishers to high-level negotiation between governments and in United Nations agencies. Despite the overwhelming economic logic that catching fewer birds meant catching more fish (or, at the very least, fishing more efficiently), engagement with those responsible for managing fisheries has been exceptionally challenging. This was particularly so with those nominally responsible for managing the relevant fisheries on the high seas, the tuna Regional Fisheries Management Organisations (RFMOs).

It was rapidly apparent that most high seas RFMOs had limited ability even to manage the target fish species for which they were established to take responsibility; commitment of time and effort to bycatch issues was seen as entirely irrelevant. Not only were most RFMOs essentially monopolistic cartels of small numbers of fishing nations with little interest beyond maximizing profit, but most were also dominated by the powerhouses of the Asian fleets (led by Japan) and the European Community (dominated by Spain). Throughout the 1990s, these two constituencies maintained an effective stranglehold over most aspects of fishing policy and practice within their domains.

It has taken a decade of work for almost all the stakeholders—from benign governments to campaigning nongovernmental organizations—slowly to bring RFMOs to start to implement more precautionary management of target fish stocks and to begin to address their international responsibilities in terms of environmentally responsible fishing, notably with respect to bycatch.

This book provides an exceptional case history of this process. At the center of the struggle is the fate of many of the iconic animals of the ocean. The species and systems of the great waters of the world belong to all of us. At present, most stakeholders have been disenfranchised by the activities of a small group of players, many of whom are in thrall to the "beneficial" owners of the overcapitalized, subsidized fleets that plow increasingly uneconomic paths across the global ocean. More transparent and democratic processes are, gradually, being developed; whether these will be enacted in time to redress the situation remains to be seen. However, for many populations of target fish species and nontarget marine mammals, turtles, and seabirds alike, it is cold comfort that wider-scale extinctions may only be averted by the economic collapse of the fisheries involved. Not only will this be the saddest of indictments of human-kind's inability to manage its affairs in sustainable ways, but it will most likely simply displace the root problem elsewhere, overwhelming the biodiversity of other fragile ecosystems. The chase for profit before sustainability, for exploita-tion of food resources without account of the damage to the environments that sustain them and the ecosystem services they provide (supporting much of the planet), places us at a critical juncture. In reality, unless new governance for the oceans can be achieved soon, based on principles of equitability and sustain-ability and with better appreciation that the world's resources are unlikely much longer to sustain our expanding population, the prospects will be bleak indeed.

This book, however, testifies to the endeavors of all those who wish to try to change the systems that we have created and inherited before it is too late. Whether there is the political will to make these changes is the key question; we have the tools and the commitment to complete the job if the systems and rules of engagement can be changed sufficiently. If so, we shall all be able to continue to enjoy the spectacle of albatrosses sailing over the global ocean; their longer-term fate, however, is inextricably bound up with our own.

Milestones

1593	Sir Richard Hawkins describes hooking huge seabirds soaring close to his vessel in the southern seas, then complains that their flesh is scant and poor compared with their impressive size.
1672	John Fryer recorded large, very long-winged seabirds he calls albatrosses on his voyage to India.
1747	George Edwards's *Natural History of Birds* depicts an albatross.
1770s	Members of James Cook's voyages mention capturing, killing, and eating albatrosses.
1798	Samuel Taylor Coleridge publishes *The Rime of the Ancient Mariner*.
ca. 1840s–1870s	Emigrants to Australia and elsewhere describe "fishing" for and shooting at albatrosses at sea. Whaling crews do the same.
1922	International Council for Bird Preservation (ICBP) is established to conserve birds, including seabirds, and collaborates with other bodies worldwide on their behalf.
1928	W. B. Alexander's *Birds of the Ocean* is the initial guide for the world's seabirds, and includes thirteen albatrosses.
1937	Eric Richdale observes and commits to protecting the nests of albatrosses close to Dunedin, New Zealand.
1947	Robert Cushman Murphy publishes *Logbook for Grace*, detailing albatross collections and studies on South Georgia in 1912–1913.
1949	The Short-tailed Albatross is incorrectly declared extinct.
1951	Seabird enthusiasts begin to study and then band (in 1958) wintering albatrosses off Queensland, Australia.

Leonard Harrison Matthews draws upon three years of work on South Georgia in *Wandering Albatross: Adventures among the Albatrosses and Petrels in the Southern Ocean.*

1955   Japanese longline fishery sets 20,000 hooks around New Zealand for Southern Bluefin Tuna on lines up to 60 miles long. A decade later, the same fleet deploys 14 million hooks annually.

1958   William Jameson publishes *The Wandering Albatross.*

1959   Twelve nations sign the Antarctic Treaty in Washington, DC, which applies south of 60°S latitude and stipulates that Antarctica should be used for peaceful purposes.

1966   International Union for the Conservation of Nature (IUCN) and ICBP publish a list of the world's endangered birds, including the Short-tailed Albatross.

1972   The United Nations Educational, Scientific and Cultural Organization (UNESCO) meeting in Paris adopts the Convention Concerning the Protection of the World Cultural and Natural Heritage, the so-called World Heritage Convention (effective December 1975). Sites inscribed by six nations are homes for albatross nest colonies.

1973   Convention on International Trade in Endangered Species of Wild Fauna and Flora (CITES) prohibits commerce in any species listed in one of three appendices. The Short-tailed Albatross appears in Appendix 1, together with five additional seabirds.

1979   Twenty-eight states sign the Convention on the Conservation of Migratory Species of Wild Animals, known as the Bonn Convention, which came into force November 1, 1983. It facilitates cooperative arrangements between range states and promotes intergenerational equity in international law.

1980   Convention on the Conservation of Antarctic Marine Living Resources (CCAMLR) concluded in Canberra, Australia, in response to fears that unregulated fishing for krill might adversely affect whales, seals, penguins, and albatrosses that feed on krill. Part of the Antarctic Treaty System, this convention with an ecosystem approach became effective in 1982.

1982	United Nations Convention on the Law of the Sea (UNCLOS), in force in 1994, concluded a so-called constitution for the oceans, providing a general scheme for authority over marine waters.
1984	ICBP publishes *Status and Conservation of the World's Seabirds* and includes the conservation status of several albatross species.
1985	Fourth Meeting of CCAMLR discusses accidental and incidental seabird mortality.
1986	Seabird Specialty group of ICBP (later BirdLife International), meeting in Kingston, Ontario, Canada, identifies "Seabirds on Islands" for conservation attention. It includes albatrosses on Amsterdam Island, the Chatham Islands, and the Galápagos.
1989	United Nations General Assembly unanimously adopts Resolution 44/225 recommending a ban on driftnet fishing. The resolution placed a moratorium on all "large-scale pelagic driftnet fishing on the high seas" by June 30, 1992.
1990	Australia presents a paper to CCAMLR meeting estimating 44,000 albatrosses killed annually by longlines outside the CCAMLR zone, and describing how collaboration with Japanese fishers using "bird-scaring streamer lines" reduces mortality.
1991	Australia ratifies the Bonn Convention and presents its first national report on albatross mortality in longline fishing based on Nigel Brothers's observations.
	Real Time Monitoring Program (RTMP) is established in Japanese tuna fishery for catch evaluation and is extended to bycatch, including seabirds, in 1995.
1992	The Mexican government organizes an intergovernmental conference that requests that the Food and Agriculture Organization (FAO) of the United Nations articulate a Code of Conduct for Responsible Fisheries. This Cancún Declaration draws upon previous provisions of the 1994 Agreement to Promote Compliance with International Conservation and Management Measures by Vessels Fishing the High Seas.
	Report of United Nations Conference on Environment and Development (UNCED) addresses bycatch problems and

promotes (under Agenda 21) practices that mitigate the take of nontarget species (similar to the Cancún Declaration). The report also stresses the need to minimize discards while alleviating incidental catch.

1993   United Nations Conference on Straddling Fish Stocks and Highly Migratory Fish Stocks begins its first substantive session (concluded in 1995).

Biodiversity Convention comes into force and includes marine ecosystems. Parties shall cooperate in high seas areas outside their jurisdictions. Enforcement is weak.

National Audubon Society (founded in 1905, now with 400,000 members [in 2010] in 500 chapters in the Americas) launches a "Living Oceans Program" to improve fisheries management. It includes longline problems and publishes a consumers' "Audubon seafood wallet card" (in 2002), with swordfish, Chilean sea bass, and Atlantic halibut in the "red" category. It supports experiments with underwater chutes for the Hawaiian tuna fishery (2003) in order to eliminate seabird bycatch.

Rosemary Gales publishes *Co-Operative Mechanisms for the Conservation of Albatross*, sponsored by Australian National Parks and Wildlife.

1994   Commission for the Conservation of Southern Bluefin Tuna (CCSBT) establishes a working group to monitor "ecologically related species," such as albatrosses, taken as incidental catch.

Australia establishes a deep-sea trawl fishery for Patagonian Toothfish around Macquarie Island. Fishery is a stepping-stone for developing the nation's Southern Ocean fishery in an eco-friendly manner.

1995   The Inaugural Albatross Conference, called the First International Workshop on Albatross-Fisheries Interactions, takes place in Hobart, Tasmania. It attracts 120 seabird and fisheries scientists, managers, industrial representatives, and conservationists from eleven nations. Participants make six recommendations about reducing albatross mortality (set lines at night, place streamers over longlines, adopt mechanical bait throwing, discharge offal on opposite side to set lines, use weighted lines, and use thawed bait).

International Conference on the Sustainable Contribution of Fisheries to Food Security is organized by Japan and the FAO in Kyoto. It focuses on the bycatch issue (although Japan regards environmentalist concerns as exaggerated and unscientific in regard to marine resources, including whales, and driftnets).

Australian Endangered Species Protection Act lists Macquarie Island's Wandering Albatross.

FAO draws up a Code of Conduct for Responsible Fisheries, which endorses a precautionary principle in fisheries management so as to minimize waste and any threat to endangered species, and promotes selective fishing methods and devices.

New Zealand introduces levies on domestic commercial fisheries to mitigate the impacts of fishing on marine wildlife, including seabirds. Levies supported observer coverage on pelagic longline vessels, improved the use of streamers, and funded efforts to reduce incidental losses.

Band returns of albatrosses taken by Japanese longline vessels reveal the global impact of fishery operations, and attendant responsibility, with a provenance of 12 islands in the Southern Hemisphere.

Linda and Warren Newton write *Floss the Wandering Albatross* (Victoria Bendigo Press), a children's book about a Macquarie Island bird.

Nigel Brothers, Parks and Wildlife, Tasmania, publishes *Catching Fish, Not Birds: A Guide to Improving Your Longline Fishing Efficiency*.

1996   IUCN Resolution on Incidental Mortality of Seabirds in Longline Fisheries adopted at the first session of the World Conservation Congress of IUCN in Montreal, Canada, calls on states and regional fisheries organizations to implement measures to reduce incidental losses of seabirds.

John Warham publishes *The Behaviour, Population Biology and Physiology of the Petrels* (Academic Press).

1997   At the fifth meeting of Parties to the Bonn Convention, Australian and Netherland delegates add albatrosses for inclusion under the convention. Appendix I, Endangered Migratory Species, adds the Amsterdam species to the Short-tailed

Albatross (listed in 1996), so that range states must prohibit the taking of these animals. Appendix II, Migratory Animals with an Unfavorable Conservation Status, requires international agreements and cooperation to manage and conserve them. Two types of agreements exist for Appendix II species: AGREEMENTS (capitalized) designed to protect birds across their entire range, and Agreements (cap/lowercase) to include species whose populations periodically cross national boundaries. Nominated albatrosses received the capitalized agreements in 1997.

The Marine Stewardship Council is founded as an independent global organization, headquartered in London, that promotes responsible, environmentally sound, socially appropriate fishing practices.

BirdLife International initiates the Global Seabird Conservation Program (with RSPB funding) with BirdLife South Africa in Cape Town. It aims to mobilize partners and sympathetic NGOs to coordinate activities and to establish a regulatory framework for reducing seabird losses.

Twenty-second session of FAO's Committee on Fisheries proposes an expert consultation for a plan of action aimed at longline seabird bycatch reduction. Japan and the United States fund the initiative; hire Nigel Brothers, John Cooper, and Svein Lokkeborg; and publish their findings as a joint report, *The Incidental Catch of Seabirds by Longline Fisheries: Worldwide Review and Technical Guidelines for Mitigation* (1999).

1998   G. Robertson and R. Gales edit *Albatross Biology and Conservation* as the outcome of the 1995 Albatross Conference (see above). They include thirteen Conference papers in twenty-three chapters.

Technical Review by BirdLife for UN-FAO of longline fisheries worldwide, as part of the basis for the FAO's International Plan of Action (IPOA-Seabirds).

EU Council of Ministers bans the use of drift gillnets for tuna in its waters, beginning in 2002.

1999   Australia's Environment Protection and Biodiversity Conservation Act (EPBC Act) protects matters of National Environmental Significance. It streamlines environmental

assessment, protects national biodiversity, and integrates
the management of important natural and cultural sites. It
also includes a Threat Abatement Plan for longlining.

2000 Second International Conference on the Biology and Conservation of Albatrosses and Other Petrels held in Honolulu,
Hawaii, May 8–12.

UK-based BirdLife International launches Save the Albatross
Campaign (as part of its 1997 Global Seabird Conservation
program) at the British Bird Watching Fair. The campaign
urges the public to pressure the fishing industry by buying
fish that is accredited as "Albatross-friendly," landed by
legal means. This NGO urges governments of longlining
nations to develop and implement National Plans of Action within the UN-FAO framework and requests them
to sign a new agreement on albatrosses, implementing its
action plan.

2001 The Agreement for the Conservation of Albatrosses and Petrels
(ACAP) opens under the Bonn Convention. This agreement is legally binding and requires signatory states to take
specific measures to reduce seabird bycatch from longlining, and to improve the conservation status of albatrosses
and petrels.

FAO adopts an International Plan of Action on Illegal,
Unregulated and Unreported, or "pirate" fishing (IPOA-
IUU). Australia had raised the issue at the Committee on
Fisheries (COFI) meeting in 1999. The topic subsequently
appeared in the Code of Conduct declaration in Rome in
March 1999 and was adopted at the FAO Ministerial Meet-
ing on Fisheries.

BirdLife International hosts a workshop in Uruguay to promote solutions for seabird losses in South America.

UN Fish Stocks Agreement goes into force as a milestone
global treaty to promote responsible fishing on the high
seas.

Audubon Living Oceans launches a Seafood Lovers Initiative, a science-based, consumer-friendly program to edu-
cate the public on how choices of seafood make a difference for ocean life.

Edward F. Melvin and Julia K. Parrish edit *Seabird Bycatch Trends, Roadblocks, and Solutions.*

2002 HRH Prince Charles hosts a reception in London for BirdLife's Save the Albatross Campaign. The Prince remains outspoken in support of the albatross.

French authorities arrest *Arvisa 1* for illegally taking Patagonian Toothfish off Kerguelen Island.

Tourists on International Association of Antarctic Tour Operators (IAATO) vessels contribute $5,500 dollars to Save the Albatross Campaign.

Audubon and its Ocean Wildlife partners are instrumental in passage of the Fishery Management Plan for Pacific Tunas, Billfish, and Sharks.

Carl Safina publishes *Eye of the Albatross.*

2003 Australians arrest *Viarsal.*

2004 The Agreement on the Conservation of Albatrosses and Petrels (ACAP became operative February 1, 2004, following ratification by five nations) holds its first meeting in Hobart, Tasmania, its current headquarters. Thirteen countries are full members (as of 2010).

Third International Albatross and Petrel Conference takes place in Montevideo, Uruguay, August 23–27.

Marine Stewardship Council (MSC) certifies South Georgia Toothfish with an ecolabel, with vigorous objections.

BirdLife publishes *Tracking Ocean Wanderers*, mapping the at-sea distribution of albatrosses through telemetry.

2005 BirdLife/RSPB's Cleo Small publishes *Regional Fisheries Management Organisations: Their Duties and Performance in Reducing Bycatch of Albatrosses and Other Species.*

BirdLife's Save the Albatross Campaign assembles an Albatross Task Force (ATF), consisting of shipboard teams that advise fishers about lowering bird deaths in longline and trawl fleets and offer onshore workshops.

THE *Albatross* AND THE *Fish*

CHAPTER 1

Storytelling

ONE AUGUST DAY IN 1998, I took a boat out of Depoe Bay in Oregon to see pelagic birds, birds that inhabit the open seas. Pelagic birds draw me, as they do a growing number of people who sail out of all three US coasts and from other ports in South America, South Africa, Australia, New Zealand, and the United Kingdom and on the ferry that links Plymouth in the United Kingdom with Santander, Spain. We wanted to see birds that remain at sea for months and years, like albatrosses, petrels, and shearwaters. Most of these species forage where the continental shelf drops off in nutrient-rich zones of currents and upwelling, and they fly thousands of miles in search of food. Some albatrosses are known to travel more than three thousand miles in a single journey to collect food for their chick.

Pigeon Guillemots scattered as our boat's muttering diesel engines nosed the thirty-foot vessel between the narrow concrete barriers of the harbor and out under the road bridge. Then, with a high-pitched whine, the skipper opened the throttle, and we sped for the one-hundred-fathom line some thirty miles west. Weather was poor the day we set out. Twenty-knot wind gusts scudded whitecaps from the tops of larger swells, making the ocean look angry. Someone shouted "Arctic Tern" as two buoyant seagulls skimmed past. The craft rolled in the choppy waters, making it hard to squint through binoculars while bracing our arms and legs on the deck. Soon, the excitement and novelty wore off, but not before the first of many torpedo-shaped Sooty Shearwaters, dressed in solid black, sculled low and fast, headed northward. A telltale glimmer of feathers or any other sort of movement among the swells drew our binoculars, but birds were hard to identify from our bucking deck. Sooties headed past, followed by Pink-footed Shearwaters showing off white bellies and whiter underwing linings as they tilted to and fro just above the surface. Then we cruised past several Northern Fulmars, each one built like a bulldog, all head and shoulders, seemingly more interested in our presence than the shearwaters had been.

We found our first albatrosses after an hour of hard driving. The crew was chumming, that is, emptying offal over the stern from cartons stowed aboard. Animal fats smoothed the wake as we slowed down, cut the engines, and drew in our first Black-footed Albatrosses, miniature B-52 bombers. Fearlessly they swung down, landed, and began gobbling chunks of meat in the spreading stain. Their voices broke the sudden silence with a medley of squawks, grunts, and brays in this middle-of-nowhere. Some showed white under-tail coverts, others seemed wholly dark, an elegant satiny chocolate brown. As the huge birds paddled to within twenty feet of our wallowing vessel, people snapped away with their cameras, catching the size, shape, and details of the plumage of these fearless visitors as they turned and twisted among the scraps. I counted twenty-seven Black-foots around us, North Pacific members of the albatross family. We pressed the rail to watch and film this minicarnival of feeding and gurgling in a wide blue-gray desert.

Without any instruments under the leaden sky and buckling waves, I hadn't the faintest idea where the coastline lay. There was nothing to satisfy the senses except the albatrosses. I felt as though I were going blind looking for some reminder of land, maybe a ship, a buoy, or the smudge of a far-off coast. Then, one after another, each bird began to patter and slowly lift off, banking to right or left with a sureness of reckoning. I had no idea where I was. Apparently, they did. I knew they would continue to know where they were and where they were going throughout their lives. I yearned after that sense of belonging and being at home. When the last bird flew out of sight, I grew even lonelier. There was nothing to relieve the silence or vast openness of the moving swells.

Oregon was the first of many pelagic trips. Over the years, I made voyages from Tasmania; out of Sydney, Wollongong, and Perth in Australia's mainland; and to the fabulous hot spot off Kaikoura on the northeast coast of New Zealand's South Island. As I imagine albatrosses do, I have relished the sense of being out there, far away from the usual hubbub, noticing ocean colors, how the water describes its own traceries and constant shifts. There is a feeling of solidity under you, bearing you up in some unfathomable way and then beckoning, come in, swim, rest here awhile.

I have observed many seabirds new to me and counted seven albatross species, half of all accepted species until the mid-1990s, when a taxonomy revision expanded the number from fourteen to twenty-four, then downsized it to twenty-one, and has now updated it to twenty-two. It is still under debate. I shall refer to twenty-two in this book for the sake of simplicity and consistency. The albatross is extraordinarily large. The largest, the group of Great Albatrosses, includes the Northern Royal Albatross, which I have witnessed gliding off Taiaroa Head near Dunedin, New Zealand, with a wingspan as wide as eleven feet six inches from tip to tip. Albatrosses are elegantly plumaged, decked out smartly in white set on black, brown, or gray, and each one carries a large bill with a menacing down-curved tip that strikes or jabs this way and that in its scrimmage for food. And yet, when they are courting, mating, and caring for their young, their faces, including the menacing beaks, become almost angelic as they curve their necks upward and spread their wings in an embrace and dance of mutual praise. In groups, they perform a gently chattering chorus of clicking and clacking greetings and reassurances.

Albatrosses are consummate aerial gliders, even in the harshest of conditions, characterized as "sail planers of stormy southern seas" by seabird scientist Ronald M. Lockley. They feed our escapist desires to be lifted from the shackles that tie us to the earth. They traverse vast distances while searching for food before turning instinctively toward their nest sites with an apparent nonchalance that belies their sureness. Someone has calculated that a single bird may make the equivalent of seven or more return trips to the moon over an average fifty-year lifespan. In the ocean wilderness that surrounds me on these pelagic trips, the albatross is utterly indifferent to my usual anxieties about location and bearing. And compared with those of the albatross, my bearings and pathways are limited, ponderous, and restricted.

Unfortunately, the albatross as a bird family is currently at greater risk for survival than any other bird group worldwide. The threat is happening in its watery stronghold, the open sea. It happens as the birds swing down to jostle over a tempting silvery flicker of fish. We've accustomed them to associate certain shapes and sizes of vessels with free meals—meals on wheels for seabirds. But often the silvery flickers of fish or squid the birds have spied are fixed on barbed steel hooks attached to the longline sliding from the boat's stern, or they are the flashing, milling fish being drawn together and up toward the water's surface in a trawl net. Instead of swooping down and lifting up again with these fish, the albatrosses become part of the catch, by the tens of thousands every year by all estimates. They are called bycatch in these instances because they aren't the fishers' targets.

There is growing alarm and concern over the depletion and collapse of fisheries as a result of global commercial operations using sophisticated technology that includes research ships, search planes and helicopters, onboard refrigeration and canning, flash freezing, and industrial longlining and trawling. The rate of damage to marine resources, despite fishery management efforts, is greatly accelerated by illegal fishing and fishing under flags of convenience from nations that do not enforce laws and regulations about fishing methods or catch limits. There are also vessels that onload illegally caught fish from sister vessels on the high seas in order to deliver these catches to the ports around the world that accept illegal shipments. There is less public awareness of the threat to entire seabird species from commercial fishing, especially longline fishing and trawling, or of our human habit of accumulating and dumping oil and other chemicals, solid waste, and plastic trash into the seas and oceans.

This book is a plea and a prayer. The plea is to take notice of this bird tragedy. Largely unseen and ignored, it is happening daily in the world's waters, especially around industrial fleets operating in the Southern Hemisphere, where most albatrosses live, but also in the Central and North Pacific Ocean. The prayer is for all to find a role in what has become, against all odds, an international environmental effort to save the albatross.

Albatrosses are naturally long-lived birds, slow to mature, usually paired for life, and parents of one chick at a time, often every other year. They cannot survive such destruction. Their populations are plummeting. BirdLife International, a major nongovernmental organization (NGO) committed to bird research and conservation, lists eighteen of the recognized twenty-two species as threatened with global extinction. Six of them are in the Endangered category, and four more are in the most serious Critically Endangered category, meaning facing an imminent risk of extinction in the wild. Eight more are listed as Vulnerable, that is, decreasing numerically toward endangerment, and the remainder are Near Threatened.[1] The good news is that the ongoing threat to seabird survival from industrial fishing can be mitigated and even prevented at minimal cost to commercial enterprises. When certain committed scientists and nonprofit conservation organizations confirmed the costs of longlining to seabirds, some vital pieces of the puzzle of falling albatross numbers began to come together. National governments, regional fishery management organizations, and commercial fishing operators, urged on by nonprofit and governmental environmental workers as well as by an increasingly informed public, began working together to lower the numbers of birds killed by the hooks. It has become clear that the fate of the albatross is linked to the fate of commercial fish. It is good news for the albatross that fishery management organizations, especially the ones with large numbers of albatrosses in their areas, are changing—or at least saying they will change—from managing a single fish species, upon which fisheries science has focused, to regarding their ocean jurisdictions as ecosystems of interdependent species.

In this book, we intend to examine an incomplete puzzle that is at the same time an economic and an environmental challenge. We can look at the pieces already identified that are slowly moving into place, hold them up from various angles to see what fits and what's still missing, and speculate on what areas of action are open to us for helping the bigger picture reveal itself and come together. On the positive side, pieces that appeared to be misfits, such as bird conservationists and commercial fishers, are discovering that in this bigger picture that is emerging they are part of the same puzzle and are, in fact, finding a fit in some areas. On the negative side, however, it has to be said that the emerging and slowly crafted fit among these pieces may be too late to save one or more albatross species and equally too late in regard to the largest commercial fish species.

The shaping of the pieces and assembly of a larger, more coherent picture concerning marine resources is not happening by itself. This story is inescapably a story of the individuals who are making it happen and how, as well as of the nonprofit environmental groups, commercial entities, and national and international government offices and agencies that are setting up a framework for conserving marine animals. Over time, these stakeholders have worked to develop the picture as it is today, not an umbrella organization as it turns out, but an intricate web of multiple overlapping authorities, scientific knowledge producers, and negotiators, and a mix of hard and soft laws that relate to the management of ocean regions and their flora and fauna. We recognize also that this particular puzzle, once it has been assembled and described, may offer a model, or at least a set of lessons learned, for other environmental regimes dedicated to the protection of endangered animals and ecosystems.

At bottom, however, it is the plight of the graceful, elegant, and long-lived albatross that propels us to tell this story. We identify with bird scientist Robert Cushman Murphy, who said, after observing his first albatross at sea in the early 1900s, that he belonged to a "higher cult of mortals, for I have seen the albatross."[2] More profoundly, we respect the albatross's being what it so remarkably is and its need to live free and beyond human interference. And this confronts us with the paradox that at this point in our history, the albatross's ability to live free and beyond human interference requires human interference—in the form of species observation, documentation, and management; successful efforts to protect remote and isolated nesting grounds; and modification of commercial practices that are currently threatening this bird group's existence. This book intends to tell—and spread—the story of a kind of interference that is collaborative, international, and potentially salvific.

The albatross is a bird most people familiar with Western literature still know as an icon of guilt, retribution, and redemption. It hung around Coleridge's ancient mariner's neck as a reminder not only of the needless death the sailor caused by gratuitously shooting it from his ship's deck but also of the destruction his needless killing brought to the entire ship and its crew. The relevance of this story to our present situation as human beings on the earth is only too obvious.

Coleridge's poem was more prescient than the author, twenty-five years old at the time, could possibly have known, although he was, with help from laudanum, also a mystic and visionary. Most people who have read the poem, perhaps with puzzlement or amusement, first encountered it in secondary school and have its stark images and rhymed rhythms imprinted permanently in their mind's eye and ear: "Water, water, every where, / And all the boards did shrink; / Water, water, every where, / Ne any drop to drink."[3] It has only been during the past fifty years that we have begun to read the visionary and spiritual aspects of the poem specifically as an allegorical warning of the potential we hold for destroying our environment entirely. The ancient mariner's only possibilities for redemption were recognizing, acknowledging, and endlessly repeating the story to everyone he met. Although narrating the story over and over was intended to set him free morally, it was too late to save the ship and its crew:

> Forthwith this frame of mine was wrench'd
> With a woeful agony,
> Which forc'd me to begin my tale
> And then it left me free.

> Since then at an uncertain hour,
> Now oftimes and now fewer,
> That anguish comes and makes me tell
> My ghastly aventure.

> . . .

> He prayeth well who loveth well,
> Both man and bird and beast.

He prayeth best who loveth best
All things both great and small:
For the dear God, who loveth us,
He made and loveth all.[4]

Writing this book puts us in the position of the ancient mariner, not only in the sense that the project has indeed at times felt like an albatross around our necks, but more often in the way we find ourselves strangely compelled to tell this story, once more, to anyone who will listen; not only to progress in our own moral redemption but even more importantly to show how individual interests and commitments, if followed truthfully and without despair, have the potential to become part of a redemptive synergy whose unpredictable outcome holds promise for the future well-being of the planet and its inhabitants.

Part One

THE ALBATROSS

CHAPTER 2

Plunder

WHEN MORIORI HUNTER-GATHERERS arrived on New Zealand's Chatham Islands chain some five hundred years ago, they subsisted on seals, fish, and birds, including nesting petrels and albatrosses. Sailing about five hundred miles due east from New Zealand, these early Maori fashioned a distinct culture on the Chathams, comprising ten islands and islets that provide a major nesting area for three of the twenty-two albatross species. The Moriori regularly harvested the seabirds by raiding nest colonies on remote wave-pounded cliffs, sculling sea canoes as far as twenty miles one way across perilous swells.

By tradition, on each trip they took about seven hundred young albatrosses at the right stage of plumpness, leaving other birds and eggs in the colonies alone, an early and prescientific example of sustainable management. They climbed the treacherous cliff faces, clubbed selected birds, and flung their bodies down to the kayaks waiting below. Once back on the main Chatham Island, they strung up the dead birds for several days to allow them to mature before stripping the meat and rendering the oil over fire. Packers trimmed the browned meat, loaded it into casks, and preserved it in the birds' own oil to take it back to the mainland in the west.

The situation changed in 1835 when Maori settlers from the Taranaki area on the North Island of New Zealand invaded the Chathams and killed and enslaved their ancestral relatives. Intermarrying with both survivors and incoming Europeans, the invaders continued the tradition of capturing albatrosses and storing their meat in rendered oil. Both men and women took to wearing albatross feathers in their pierced ears and turned sprays of white albatross down into ornamental flowers called *pohio* for hair adornments. They fashioned pendants from the seabird's white body down and carved bird bones into awls, needles, fishhooks, toggles, flutes, and small fifes. They also used the birds' bones to carve ornate tattoos on their bodies and decorated their canoes with white albatross feathers to signify swiftness. The custom of preserving the heads of both relatives and enemies included dressing a trussed knot of hair with albatross feathers.

The Maori began to ship albatross meat and feathers to Taranaki, their original home, partly as an offering to Te Whiti, their great spiritual leader who lived there. *Pohio* later became the symbol for the Te Whiti movement, the Taranaki pacifist movement begun by Te Whiti-o-Rongomai and his nationalist brethren in the late 1870s. Right of conquest and native land titles over bird islands found favor in New Zealand's native land courts in 1885 and 1887, although laws failed to recognize Moriori first rights.

After 1970, as seabird conservation and conservation in general became a worldwide concern, the Moriori became concerned with protecting their traditional rights to island birds as food. In 1989, applications to the Ministry of Conservation sought legal standing for taking albatrosses on the Chathams as traditional meat as well as for their feathers for mattress and pillow down and their oil for medications. As the New Zealand government has worked to protect and manage albatross nesting islands in modern times, the Maori have tried to reserve access to the breeding sites for this ethnic staple, but the New Zealand Parliament has yet to codify this right in statute.[1]

When European explorers and sailors began in the sixteenth century to extend their reach into the Southern Ocean, more than two hundred years before Coleridge's "ancient Mariner" killed an albatross, they had no idea that their entry into the albatross's ecosystem would wreak such destruction on Southern Ocean flora and fauna. As it turned out, the fate of the albatross was tied to the fate of fur and elephant seals and the whales of the region for whose oil, meat, body parts, and fur consumers in the Northern Hemisphere were willing to pay premium prices.

It was then that Europeans encountered albatrosses for the first time, flying over the open seas or nesting on some of the islets and rock eruptions that formed archipelagoes in the Southern Hemisphere. The letters and journals of these men record their amazement at seeing these soaring goliaths. Unlike Coleridge's ancient mariner, crewmen often took albatrosses to be good omens. They were birds of mystery, appearing suddenly from nowhere in the wide expanses of water and wind. Some seamen thought they were the souls of shipmates arising suddenly from the rolling ocean itself. Other sailors regarded them as companions in a lonely, wild, often boring, and cramped ship's world on the open sea. One vessel's master enjoyed an albatross as a tilting and tracking colleague for six successive days, and some officers reported that the same bird could track a ship for hundreds and even thousands of miles, in one instance for almost three thousand miles.

The mariners recorded various methods for snagging albatrosses by fishing for them and later by shooting them with firearms. They paid out a line with bent-nail hooks hidden in pork or other tidbits in the ship's wake, hooked the scavenging albatross as it swooped down to take the bait, and hauled in the struggling prize for the pot, fresh meat for hungry men. Joseph Banks, a naturalist on Captain James Cook's ship *Endeavour* from 1768 to 1771 on his first circumnavigation of the globe, reported February 5, 1769, "myself a little better than yesterday, well enough to eat part of the Albatrosses shot on the third, which were so good that every body commended and Eat heartily of them tho there was fresh pork upon the table."[2] Officers and men continued to take albatross meat on Cook's subsequent two voyages into the Southern Hemisphere, the last of which ended tragically in 1779.

Some shipmen treated the docile birds as pets, hobbling them but allowing them to roam their vessel's deck. They teased them; mimicked their ponderous, wobbly gaits; and occasionally took pleasure in torturing their charges. Rarely did they release them, although some birds found freedom after being tagged. In Herman Melville's *Moby-Dick* (1851), the sailors loosed an albatross with a marked leather tag. In 1886, an Australian newspaper reported that a boy in Freemantle had come across a recently dead albatross with a message on a tin band around its neck. Written in French, the message reported that thirteen shipwrecked sailors were marooned on the Iles Crozet, thirty-five hundred miles west of Australia. A search party set out, reaching Crozet about two months later, only to find a second message that indicated the party had headed for a neighboring island. The sailors were never heard from again.

People traveling and emigrating to Australia and New Zealand also passed the time catching albatrosses. They fished for them and later on shot the "elegant birds gracefully hovering about our rear." Some killed them for sport, wagering on marksmanship and the number they could down in the wake. "It was a grievous sight to see the splendid bird float off, dead,"[3] lamented one passenger. Sailing vessels replaced by reliable steamers ferried millions of emigrants from Europe into the Southern Ocean, over one million to Australia from 1788 to

1880, for example. Records suggest that thousands of seabirds died during these voyages, especially the Wandering and Black-browed Albatrosses that swung behind vessels after rounding Africa's Cape of Good Hope, where astronomer William Bayly on Cook's second world voyage (1772–1774) declared that "they are easily caught."[4]

A source of long-lasting and ongoing devastation of seabird colonies was the introduction of European livestock and other alien species of plants and animals. Nonnative plants and animals, especially invasive plants and predatory animals, devastated nest colonies. Cattle, sheep, and goats; household pets, including dogs, cats, and rabbits; and the inevitable ships' mice and rats altered these remote, isolated places through habitat transformation and animal disturbance, destruction, and predation. Livestock consumed tussock grasses and other vegetation that albatrosses pulled to build and maintain their pedestal nests. Cats and rats preyed on the nesting birds, especially helpless chicks. The same pattern of habitat change and predation from introduced alien species occurred on all the islands suitable for human settlement in the temperate and colder latitudes of the Southern Hemisphere, notably Tristan da Cunha, South Georgia, and the Falkland Islands claimed by Britain; Kerguelen and Amsterdam Island claimed by France; Macquarie Island claimed by Australia; and possessions of New Zealand such as the Auckland Islands, Campbell Island, and the Chatham Islands.

The most grievous sustained injury to the albatross happened during the sealing and whaling era that started in about 1780 and persisted through World War I. As hunting vessels moved poleward in the Southern Hemisphere, early adventurers "discovered" and "named" irregularly spaced sets of remote, often fog-draped and ice-covered volcanic archipelagoes that circle the higher southern latitudes. Many are the exposed tips of the underwater ridges that define the shifting and collision of the earth's drifting plates. Venture capitalists read the explorers' maps and journals and eagerly bankrolled expeditions to sail out of northern ports to harvest the fur seals, elephant seals, and whales that reportedly lolled and breached about these island specks in staggeringly high numbers. Early industrialists also outfitted voyages to dig out layers of the bird guano that accumulated in seabird nest colonies to be hauled off and sold as fertilizer. As the roughneck sealing and whaling gangs headed into these new treasure troves, they also noticed albatrosses gliding beside and behind their vessels and devised various methods for catching them. Once ashore, they found albatrosses even more to their liking, especially their eggs and young chicks, developing a taste also for elephant seal tongues and male whale sexual organs. The adult albatrosses sitting on their eggs and the downy young chicks clustered in the headlands were easily approachable, and hungry men looking for something to spice up their ship's fare eagerly carried off both eggs and chicks. A market also quickly developed for albatross eggs among collectors and for writing pens made of albatross wing pinions.

Hunters walked up to the docile birds and clubbed them for food and for barter with other crews. They also fashioned their bones, skin, and webbed feet into pipe stems, tobacco pouches, and foot warmers, and sold bird skins and the large white albatross eggs to middlemen on UK and US docks.[5] For example, on the Falkland Islands, reclaimed from the Argentine government by Britain in 1833, American sealers and Argentine residents established an annual tradition of gathering albatross and penguin eggs on November 9, Lord Mayor's Day in the United Kingdom and a school holiday in the Falklands. On that day, children scurried to unload seabird eggs that workers had brought by the boatload into Stanley from the outlying islands.

Eight hundred miles southeast of the Falklands, whalers and sealers also ransacked Wandering Albatross nests on South Georgia. Sealing began there in 1786, and commercial whaling, at the close of the nineteenth century, led by Norwegian entrepreneurs. During their killing and butchering, hungry crews roamed the coastline, including those of the offshore islands, and plundered colonies of the largest birds among the twenty-seven nesting bird species on South Georgia. Nearly five thousand miles away and on the other side of the Antarctic continent from South Georgia, beach caves on Australia's Macquarie Island have yielded the bones of scores of Wandering Albatrosses killed and eaten by sealers, to the shocking extent that by 1913, only one live pair inhabited that sub-Antarctic island. Similar killing took place in Tasmania, where that state's Shy Albatross suffered terribly from human persecution.

Threat processes are dynamic and change over time. Comparing the irregular records from journals of explorers, ships' officers, and passengers with more standardized notes of early naturalists and collectors and with more up-to-date scientific observations gives ornithologists and other field workers a sense of the extent of damage done during the periods of exploration, commercial exploitation, and settlement in the world's oceans where albatrosses nest and forage. The 2010 Red List of Threatened species published by the World Conservation Union, formerly known as the International Union for the Conservation of Nature (and still often referred to as IUCN), suggests that the damage was severe.[6]

To supply the needs of an industrializing economy, Europeans and Americans wiped out entire populations of marine animals in a boom-and-bust pattern that still goes on in today's commercial fisheries. After extirpating Antarctic and additional fur seal species, sealers turned to southern elephant seals and then to whales for oil. They used their ships to double as whalers, chasing right whales along the colder coasts and sperm whales in warmer waters en route both to and from the Southern Hemisphere islands. Sealing opened up the remotest places in the sub-Antarctic, and elephant sealers hit almost all the albatross islands, including the major breeding sites on South Georgia and the Falklands in the South Atlantic; Tristan da Cunha farther north in the Atlantic; eastward to Prince Edward, Crozet, Kerguelen, and Amsterdam in the Indian Ocean; down to Macquarie and into the southwest Pacific Ocean around New Zealand. In addition to blood-drenched rocks and beaches, the gangs left other more indelible marks: a cargo of alien plants and animals that posed ongoing problems for indigenous species and the governments that decided to restore the island killing sites. On South Georgia, ships' men had begun to carve into breeding elephant seals by the 1820s and, over the following fifty years or so, reduced populations everywhere to what scientific observer Charles Leonard Harrison characterized as a "pitiable remnant."[7] On the frigid and most remote Heard Island, south of Kerguelen in the Indian Ocean, sealers lived for as many as three years at a time in exposed, filthy huts where winter temperatures hovered around freezing, subsisting on local wildlife as they hacked this elephant seal fishery.

This was the pattern everywhere as an onslaught of landings on each new site cleaned out the stocks of skin-producing animals. Then as elephant seal numbers also crashed, hunters hit the shore colonies of penguins for their oil. Barely twenty years into it, as early as 1818, ships' captains and crews complained that wherever one sailed, animal populations had collapsed or were dropping off quickly, and their underwriters found it increasingly uneconomic to go literally to the ends of the earth to finish off a remnant stock of marine beasts.

As a result, the great era of marine mammal and bird plunder in the Southern Hemisphere had dwindled and then for the most part ended by the 1930s and '40s. In fact, World War II helped bring matters to a head. World War I actually increased the demand for whale oil, from South Georgia for example, to be processed into glycerol for nitroglycerine to be used in ammunition. This brought more whale catchers into the Southern Ocean. But during World War II, other chemicals replaced whale oil, and only the Grytviken whaling station on South Georgia operated throughout the hostilities. Today, reminders of the callous brutality toward mammals and birds in the sealing and whaling eras exist in the skeletons of factories and bones of animals strewn along the coast of South Georgia. After World War II, however, for reasons we shall note, governments began restoring and managing key albatross breeding areas, in many cases designating bird islands as preserves and sites of special national and even world importance.

But after the mid-1800s, albatrosses encountered an additional devastating threat in both the Southern and Northern Hemispheres. Millinery interests in Europe and North America began to promote wild-bird feathers as ornaments in fashionable dress such as feather boas, fans, and wavy breeding plumes called "ospreys" on bonnets and hats. Some hats consisted entirely of bird feathers. Newly established fashion magazines promoted chic wild-bird-feather styles among middle-class ladies, who eagerly adopted spring and fall fashion fads that included the body parts and plumage of dozens of bird species, from tiny hummingbirds to the huge albatrosses and other seabirds such as gulls and terns. Agents representing the three major centers for haute couture, Paris, London, and New York, paid farmers and fishers working in the desired birds' nesting areas to procure as much of the novelty trim as possible. Local farmers and fishers shot herons, seabirds, and additional water bird species in their rookeries, preferably when they sported colorful and showy nesting plumage, first along shores of the United States and Central America. As more and more ships went to the Southern Hemisphere on commercial ventures, millinery and clothing agents expanded their domain into Asia and Australasia in order to satisfy the burgeoning market for decorous feathers and softer body down.[8]

North Pacific islands have been more welcoming to explorers, mercenaries, visitors, and settlers than Southern Hemisphere islands. Those islands exemplify an even broader range of human activities that affect albatrosses, and North Pacific albatrosses have been savagely attacked on their nest islands, probably more so than their Southern Ocean counterparts. Lisianski Island in the Hawaiian Island chain, about nine hundred miles northwest of Honolulu in the north-central Pacific Ocean, was an early bird-specimen collecting site for Russian seamen who hunted for guano (but failed to find it there). Then, when the millinery fad for wild feathers peaked, scores of Japanese hunters landed on Lisianski in 1904 and decimated breeding Laysan and Black-footed Albatrosses before US authorities, alerted to the killing, arrested them. In less than six months, millinery men reportedly had slaughtered almost 285,000 albatrosses, snapping off wings to pluck out rakish primary feathers to provide a "Robin Hood" look on fashionable hats. They skinned the fine breast down to make toiletries like powder puffs. Five years later, another band of feather workers revisited Lisianski and also put ashore on Laysan Island, some 115 miles to the southeast. They slaughtered an additional seventy thousand seabirds and gathered up more than a ton of albatross feathers valued at almost $100,000.

Earlier, in the 1890s, Laysan Island had been a valuable source of guano. In 1908, the island's "owner," Max Schlemmer, a foreman of the guano extraction company, had taken over operations after the fertilizer outfit moved out. He leased rights to a Tokyo-based firm to also remove and sell "products of whatever nature"[9] from Laysan and Lisianski Islands. Subsequently, gangs on Laysan harvested more than a ton of seabird, mostly albatross, feathers and seventy bales of wings for shipment, storing the plumage of thousands upon thousands of birds in the abandoned guano sheds. They ransacked Lisianski as well. Backed by entrepreneurs and millinery agents, the slaughter on these remote mid-ocean islands and Midway Atoll even farther west was so comprehensive that people feared the birds would be declared extinct. By the early 1900s, the lucrative trade in wild-bird feathers from breeding colonies in the central-west Pacific Ocean peaked at several hundred tons annually. When word of carnage on US-controlled areas filtered to Honolulu, officials dispatched a naval vessel whose officers discovered bales and bags of albatross wings and body feathers, eggs stored in barrels, and mutilated rotting carcasses. This military intervention gathered plumage valued at $130,000 but failed to halt persecution on Laysan and Lisianski. Seabird killing continued sporadically until the demand for feather millinery subsided during World War I.

The United States made claim to Midway Atoll under the Guano Islands Act of 1856, and in 1902, President Theodore Roosevelt protested to the Japanese government about "fowling" carried out there by its nationals and made the US Navy custodian of the islands. The United States used Midway as a base for the first trans-Pacific telegraph cable laid a year later, and the president and the US Congress set aside the entire atoll chain except Midway as the Hawaiian Islands Bird Reservation.

A new status did not prevent additional waves of destruction for the Laysan Albatross. First, in 1935, Pan American Airways selected Midway as a stopover for Flying Clippers en route from San Francisco to Macao and touted the nesting albatrosses as a viewing curiosity for the well-heeled passengers. Then rats arrived about 1943, and since no resident was allowed to keep house cats, the rats had the run of the place, nibbling on both eggs and chicks of the smaller seabirds. Then as the United States entered World War II, the military began expanding its presence, engineering massive earth removal, enlarging runways, adding housing and office buildings, permanently changing and in some cases destroying albatross nest habitat. Construction crews literally ran over sitting birds, bulldozers and graders crushed the fearless avians, and airplanes regularly collided with them, resulting in military orders to get rid of the "pesky birds."

By 1945, approximately fifteen thousand people lived on Midway. "Gooneys," as military personnel named the big, docile-looking birds, lived there, too, still faithfully returning to their nest sites. Black-footed Albatrosses stuck to the beach fringes, mostly away from human beings, and proved feisty and unfriendly. Laysans, however, preferred the interior, taking a shine to lawns, gardens, and other open areas around sheds, hangars, and military houses. They wandered under new shrubs and trees like peacocks in some potentate's ornamental garden. The gooneys lived up to their pet name by dancing, stepping, and braying during socialization and courtship and gradually endeared themselves to personnel, especially children.[10]

Then, in the 1960s, the US military constructed a forest on Eastern Island of three-hundred-foot towers, cables, and guy wires for a PACSCATTER apparatus, a relay system by which the military communicated with the outside world. Until it was removed, the web of steel cables proved to be an especially effective bird killer and maimer. With their long wings and large bodies, albatrosses were unable to take quick evasive action, and they fell by the hundreds—over four hundred died in one gale-force storm by being blown into the high-slung apparatus.

Additional islands in the North Pacific Ocean reflect our heavy hand on albatrosses. One of them, called Torishima, about three hundred miles south of Tokyo, is a 1.6-mile-wide island with an active volcano at its center. It is the most southerly but for one of the Seven Islands of Izu, and rises to thirteen hundred feet above sea level. The major breeding site for the delicately colored white and black Short-tailed or Steller's Albatross, Torishima is also a memorial to a great butchery of this albatross. This Northern Hemisphere seabird cruised the western Pacific Ocean, including Alaskan waters, breeding on small islands around Japan. It also commonly soared northward into the Aleutian chain, and records from Native American coastal middens place short-tails as far south as Southern California islands. Today, however, because of prolonged and intense persecution on its major breeding site of Torishima, the Short-tailed Albatross is a rare but recovering species. Persecution extirpated it from several other western Pacific islands, including Bonin, where efforts are now under way to restore it.

Torishima stands in a "typhoon alley" where high winds and rains bite into steep lava-strewn slopes around the central caldera. A massive eruption in August 1902 slung ash and lava everywhere, killing 129 residents of Tamaoki village and displacing the relict population of albatrosses. During the millinery boom, gangs killed albatrosses on Torishima as elsewhere, stripping hundreds of thousands of birds and rendering carcasses into oil and fertilizer. One estimate suggests that in less than twenty years after 1887, when the first permanent human settlement was established, reportedly five million Short-tailed Albatrosses died on Torishima. Despite edicts against fowling, slaughter continued off and on through the late 1930s. By that time, perhaps only thirty to fifty birds survived. Actually no one knows when the killing stopped, but by the late 1940s, one expert judged—mistakenly as it turned out—the Short-tailed Albatross to be extinct.

All of these relatively sudden and intense human invasions into the previously remote and secure albatross ecosystems in the southern Atlantic Ocean, the Indian Ocean, and the Pacific Ocean, both South and North, had a severe effect on albatross populations. Colonies of several species declined by huge numbers. In addition to the habitat transformation and animal destruction by human beings, the natural dynamics of the oceans surrounding albatross nest islands plus the complex relationships with longer cycles of weather and climate, such as shifts in El Niño in the Southern Hemisphere, resulted in significant fluctuations in the quantity and type of marine food available to albatrosses and their seabird cousins, which also produced shifts in seabird numbers. Because albatrosses and other seabirds as well as many Antarctic fish species are slow to mature and reproduce, have low birth rates, and are long lived—known as K-selected species—the arrival and destructive participation of humans in the albatross ecosystem, added to existing natural climate oscillations, undoubtedly tipped the survival balance for a number of species.

CHAPTER 3

Science

EARLY SYSTEMATIC ATTEMPTS to observe and understand the albatross took place before organized seabird taxonomy existed, and before many of the birds we know today had even been identified, much less understood in terms of ecology and life histories. As early as 1593, Sir Richard Hawkins, Elizabethan adventurer, buccaneer, and Member of Parliament who made the passage around Cape Horn to harass Spanish vessels, described birds that probably belonged to the species we call the Great Albatross as spreading their wings almost two fathoms. In describing seabirds new to them, English mariners used the names of common and familiar birds—swan, eagle, hawk—while Spanish and Portuguese mariners used versions of Spanish names for huge birds, such as *alcatraz* derived from *al-cadous*, the Arabic name for pelican. In a 1719–1722 navigation of the globe upon which Coleridge reportedly based his ancient mariner's voyage, Captain George Shelvocke refers to the paucity of marine life except for a disconsolate black "albitross." Joseph Banks referred to "Albatross or Alcatrace" as late as 1769 and thereafter used the term "albatross."[1] The large-format illustrations in George Edwards's 1747 *Natural History of [Uncommon] Birds* includes an early illustration of the albatross (plate 88).

Textbooks characterize the early 1900s as the heroic era of polar expeditions led by Robert E. Peary, Roald Amundsen, Robert Scott, Ernest Shackleton, and Douglas Mawson, and these derring-do men captured the imagination of Europeans, Americans, and Australians. For some, these adventures created a thirst for firsthand studies of the flora and fauna in these extreme landscapes, bitter deserts of ice at the edges of the habitable world. Museum owners began to contract with ship owners to take scientists on board, and some museums set up their own expeditions to observe and collect birds. But species observation and population estimation in that part of the world did not begin in any systematic way until the early years of the twentieth century.

We get a sense of the amazement with which early collector-scientists first encountered the albatross and then began to study it from the notes and publications of Robert Cushman Murphy, Leonard Harrison Matthews, W. B. Alexander, and Lancelot Eric Richdale. At the time, none of them knew where their investigations would lead. They could only hope their work would become part of a larger body of knowledge, in this case about seabirds and especially albatrosses. As it turns out, it has. One of the basic themes running through this story is the essential contribution the producers of reliable knowledge and data make to all efforts to understand, manage, and protect natural resources. Gathering accurate information in the tempestuous conditions of the Southern Ocean and its sub-Antarctic islands has been and continues to be extremely challenging, something to bear in mind when reading about the accomplishments of these early albatross scientists.

A little less than a year after he graduated from Brown University, and a few weeks after his wedding, twenty-five-year-old budding bird scientist and collector Robert Cushman Murphy set sail in 1912 for South Georgia as an assistant navigator aboard the New Bedford–based whaling brig *Daisy*. Urged by his fiancée, Grace Emeline Barstow, to take this golden opportunity to launch a professional career, Murphy had moved up his wedding date and then steamed off with his new bride to the Lesser Antilles to catch up with the whaler. The lanky Long Islander's job to observe and collect animals was the result of an agreement between the *Daisy*'s owners and Murphy's friend and mentor Frederic A. Lucas, director of New York's American Museum of Natural History. Lucas tasked Murphy, as a representative of both the American Museum and the Museum of the Brooklyn Institute of Arts and Sciences, to bring back bird specimens from South Georgia and to secure skins and skeletons of the elephant seal. During the *Daisy*'s commercial cruise for elephant seal and sperm whale oil in the South Atlantic, Murphy bunked in the officers' quarters with the "Old Man," Captain B. D. Cleveland out of Martha's Vineyard, who had explored South Georgian waters earlier aboard the same ship.

Murphy considered the *Daisy* to be a profitable vessel. Launched forty years before he set sail in her, she was the largest of thirty or so wooden-hulled vessels built on Long Island. A New Bedford group had purchased her in 1907 and refitted her as a whaler. Hand-framed in white oak and chestnut with planks cut from yellow pine, the *Daisy* was in fact a living remnant from a New England fleet that decades earlier had numbered more than seven hundred vessels. She was the last whale ship and hers the last crew to visit South Georgia under sail in order to pursue the perilous task of harpooning whales from open rowboats.

Nothing thrilled Murphy more than his first encounter with the "long-anticipated" Wanderering Albatross as the *Daisy* was pressing south parallel with Rio de Janeiro. He said it was one of those rare moments when reality actually transcended his expectations. The big black and white albatross "appeared merely to follow its pinkish bill at random," turning and tilting in the "invisible currents of the breeze" that made it "even more majestic, more supreme in its element than my imagination had pictured."[2] At times, the huge seabird barely hung suspended above the ship's deck with wings that "seemed to be neither beating nor scarcely quivering."[3] At other times, the adult-plumaged male slipped astern to trace lazy loops back and forth across the ship's wake. On one occasion, the handsome bird turned so sharply that one wingtip sliced the water's surface. It landed twice and appeared as "gigantic as the iron birds on the swan boats of my boyhood memories," before slapping its large feet a hundred paces in a taxi to get airborne again.[4]

Less than a month later, Murphy enjoyed his first glimpse of South Georgia. On November 24, 1912, as a whaler eased the *Daisy* to her anchorage, Murphy described South Georgia's serrated backbone of snow-covered peaks as "the Alps in mid-ocean."[5] Hundreds of Wandering Albatrosses "towered" over the sun-filled Cumberland Bay like hawks in thermals. They "circled far up under the blue vault, until they became almost indistinguishable specks against the light cirrus clouds," he noted, estimating them to be as high as perhaps five thousand feet above the sea surface. Other seabirds, including petrels and smaller albatrosses, whirled with them.[6]

As assistant navigator, Murphy worked with the thirty-four-man crew manning oars in one of the brig's open whaleboats. He rode in a steam-driven Norwegian catcher with a harpoon gun on its bow as it chased spouts the Old Man decided were good targets. In mid-ocean, Murphy shot seabirds from the brig's dory for their skins, and on South Georgia, he rigged a camp from which he studied breeding albatrosses and other seabirds as well as the endemic land-dwelling pipit.

Gangs of men commonly put ashore on nest islands for fresh albatross eggs and also for well-grown plump chicks. Gastronomically speaking, Murphy judged albatross eggs both "a delusion and a snare." After boiling the large, one-pound but rather fragile egg and supplementing its snipped end with table salt, Murphy found the first taste delicious. But as one's spoon sank lower into the yolk, the gourmand's pace slackened, and one wished for "a somewhat smaller egg." He admitted, however, that after a few days, "you once more fall a victim to the insidious temptation that never fails to cloy."[7]

Murphy grew into a careful and accomplished fieldworker, making copious bird-related entries in his journal, proudly modeled after journals Darwin penned on the *Beagle*. He discovered quickly that preserving specimens in cramped salt-laden spaces in stormy seas and then on remote, frigid island hill slopes was physically demanding. He went to great lengths to obtain skins, protecting them from saltwater and spilled chemicals. The pelagic birds Murphy was seeing in great abundance were rare in New York collections, and his 1912 expedition sponsors wanted specimens for their own collections as well as for public viewing in museums in New York. He assembled a series of skins that established the initial zoogeography for many of the little-known Procellariiformes, or storm birds, from *procella* for "storm," the order that includes albatrosses.

With support from Captain Cleveland and other officers, Murphy began to trail baited hooks behind the *Daisy* in order to lure seabirds down from the air. After they had pounced on the trailing baits and were flying off with them, they looked like "animated kites."[8] He refined his bird fishing and snagged a dozen Wandering Albatrosses in one session. He banded and released eight and made specimens of the remaining four. A week later he lowered the dory *Grace Emeline* to shoot additional birds. Undisturbed by gunfire, seabirds sailed close by, gliding along the bulging gray swells until he and the ship's cooper, Jose Correia, who on subsequent voyages also collected specimens for the American Natural History Museum, had shot thirty-six, including several albatrosses, despite grumbles from Captain Cleveland. "It appears that this is still taboo," Murphy noted, "whether the weapon be a crossbow or shotgun." "No curse attaches" to the crew's fishing for albatrosses for making pipe stems, needle cases, pouches, or even a tasty stew, "but shooting one is different."[9] Nevertheless, Murphy continued to improve his marksmanship and learned to shoot so that his quarry fell on the deck or within easy sculling distance. One Sooty Albatross he downed from a squadron of four that had been accompanying the ship for eleven days "came hurtling down into the Old Man's open arms."[10] The remaining trio of elegant brown birds continued to visit the *Daisy* for another three days. When he shot a number of recently fledged Wandering Albatrosses some nine hundred or more miles from their birthplace on South Georgia, he concluded, incorrectly as it turned out, that albatross parents starved their chicks into leaving the nest.

Murphy wrote about a thousand words a day about duties on the brig, activities with the crew, and the various seabirds he was observing and learning to identify. He noted the efficiency with which albatrosses flew, their greetings of one another on the water, their weathering of harsh storm winds and the long distances they appeared to be traveling from their nest colonies. Once ashore, he focused on nesting albatrosses, which to him looked like white sheep scattered across South Georgia's open slopes. The large numbers of Wandering Albatrosses on and around the whaling grounds of South Georgia impressed him and corroborated earlier accounts of massive populations in this sector of the Southern Ocean. He noted plumage, courtship, nest construction, incubation, diet, and feeding and correctly concluded that once the young birds left their nests, they did not hang about but put out to sea.

What Murphy reported and collected for his museum mentors during the yearlong journey launched his substantial and enduring reputation as a field naturalist. Eventually, he supplied his sponsors with five hundred bird specimens from fifty-five species and one hundred sets of wild-bird eggs as well as samples from many other mammals, invertebrates, fish, plants, and rocks. Material from his South Georgia expedition appeared in sixty-seven articles and most importantly formed the basis of his acclaimed 1936 *Oceanic Birds of South America.* Murphy drew from his carefully annotated journals in writing this two-volume book as well as from a century's worth of published observations made by colleagues on earlier expeditions. As a result, his work laid a solid foundation for the next generation of scientists, such as Lance Tickell in the early 1960s and later researchers with the British Antarctic Survey on South Georgia.

Significantly, Murphy expressed concern in his journals about the costs of the wholesale slaughter of marine animals in the Southern Ocean, especially for albatrosses. He speculated that both European whalers and American sealers, including mates on the *Daisy*, were beating down the numbers of albatrosses and other edible seabirds, causing him apprehension for their long-term viability. The effects of egg collecting on the birds also troubled him. In season, whalers from the land stations sometimes gathered "more than a thousand eggs a day,"[11] that is, a thousand eggs taken from albatross pairs that laid only a single egg every one or two years. The absence of sensitivity toward conserving the marine animals he was seeing for the first time depressed him.

Murphy noted a problem of a different nature around the major Grytviken oil-rendering station on South Georgia. On a late-November hunt for humpbacks some thirty-five miles east of Grytviken aboard the steam-powered whale catcher *Fortuna*, he came upon "more birds than one could believe existed on earth."[12] Fog had rolled in that day, not only stilling the drumbeat of harpoon guns but also grounding "millions and millions of petrels and albatrosses" on the water's surface while other birds circled low "like snowflakes"[13] filtering onto these masses. The *Fortuna* chugged through this vast bird assembly that spread across the stilling waters, bumping aside indignant-looking albatrosses. These seabirds had learned to accompany the inflated whale carcasses as the catchers towed them into port, squabbling over blood clots and the crustaceans that the dying leviathans belched up. They lunged for the fibrous circles the crew cut through the whale flukes in order to chain and tow the dead mammals. Murphy noticed albatrosses even tugging at the plugs inserted into the whales' blowholes to hold in compressed air to float the beasts. The string of whale stations and their surroundings also supplied free food for seabirds as gangs flensed and cut up the huge animals.

Scenes of whaling drawing in milling seabirds led Murphy to believe that thirty years of intensive marine mammal slaughter in southern waters had unbalanced the feeding habits of scavenger birds and had most likely increased overall numbers, in the short term, by making offal so readily available. Albatrosses must indeed have prospered from the ample waste meat and oil the sealers, whalers, and factory stations discarded. However, as winter set in and the hunting season ended, he began to notice that the birds hanging around the factories were dying, likely starving to death. Reliance on easy pickings had caused the birds to abandon their foraging instincts, Murphy feared.

Murphy reported another "canker" in this "Garden of Eden."[14] The Norway or Brown Rat had scuttled ashore decades earlier from sealing vessels and progressively gnawed its way into colonies of nesting terns and petrels, turning live birds into "vast graveyards of clean-picked bones"[15] that he found heaped around Cumberland Bay, the main landing place. His observation foretold the devastation of albatross nesting colonies on their Southern Ocean nesting islands caused by the arrival and proliferation of alien species of animals and plants that destroyed their habitat as well as eggs and chicks.

Not many years after Murphy's visit, British biologist Leonard Harrison Matthews spent several summers working with a Norwegian whaling and sealing gang on South Georgia. The numbers of birds drawn to shore-based whaling stations and floating factories astonished him, as did the birds' incredible tameness. Black-browed Albatrosses dismayed Matthews by begging like farmyard ducks around the Husvik Whaling Station in the 1926–1927 season. Food was abundant, so easy to find that he judged pickings around the rendering and freezing plants up and down the coast had led to an overpopulation of seabirds. Butchering operations attracted so many scavengers that Matthews decided, as Murphy had before him, that the birds' migratory instincts and behavior had been dulled. Even his favorite Wandering Albatrosses seemed disinclined to wander as much as he considered usual, and as a result, large numbers of seabirds starved to death after whaling closed for the winter.[16]

Like Murphy, Matthews found albatross eggs and birds delicious. Both adults and young were good to eat, the chicks served up "as white and tender meat."[17] He especially liked the ship's cook's masterly presentation of a deep bed of onions and Argentinean beans swimming in fat "with braised albatross chunks nestling as comfortably on it as they had in their tussock cradles a few days before. Delicious."[18] Matthews beached a small wooden blunt-nosed pram on the northwest end of Bird Island and filled it "to the point of sinking"[19] with tasty Wandering Albatross eggs. The contents of one egg filled a frying pan. But he found the egg collecting excessive and worried about the effects of nest robbing on bird populations. Despite his doubts, Matthews helped transport two thousand eggs from a nest colony to his vessel over a three-day period in January 1926. Unlike the docile Mollymawk Albatrosses, which gave up without much fuss, and also

Grytviken Whaling Station was a port for Murphy, Matthews, and others working on South Georgia. The hulks of two whale catchers reflect the era of marine mammal and bird exploitation. Courtesy Greg Lasley.

27

suffered from egg pilfering, a Wandering Albatross had to be literally booted off its two-foot-high pedestal lest the crewmember receive a severe stabbing as he slid his hand under the sitting bird to steal its egg.

Like Murphy, Matthews inventoried the plants and animals on and around South Georgia and worked ashore under a flimsy canvas to perfect data collection and recording. He grew interested in the reliance of native people and local settlers on seabirds for food. Although he was taking part in egg and chick harvesting, he lamented how Europeans seemed to want to exploit anything for a profit and eventually became convinced that the killing of birds and mammals should end.

While working on South Georgia, Matthews also served as a staff member at Discovery House, opened in 1925 on King Edward Point near Grytviken Whaling Station and named after the steam-assisted sailing vessel of Antarctic expedition leader and national hero Captain Robert Falcon Scott.[20] Funded by taxes on the whaling industry, the British Colonial Office, under the auspices of the Discovery Committee, had launched a series of scientific expeditions and shore-based studies. Scientists like Matthews counted whales to get a better sense of how many could be harvested, charted south polar currents, and inventoried marine organisms.

For his work on the physiology of reproduction in marine animals in the Southern Ocean and for additional research on later expeditions to Africa, Matthews was appointed scientific director of London's Zoological Society. In 1954, he was elected Fellow of the United Kingdom's Royal Society, that nation's academy of science. Colleagues lauded him for the originality of his research and the beauty of its presentation. In writing his books and articles, including the 1951 book *Wandering Albatross*, Matthews drew on his personal experience and observations as well as on earlier records and publications, including those of Robert Cushman Murphy, assembling all available knowledge about these birds for subsequent researchers.

In the 1920s, British-born and Cambridge-educated Wilfred Backhouse Alexander, known as WB, was making prolonged ocean voyages connected with his research in Australia on how to deal with prickly pear infestations, an introduced cactus that was spreading quickly in eastern Australia. During voyages to South America on behalf of the Australian Prickly Pear Board, Alexander took notes on and made detailed sketches of the seabirds he observed at sea. Planning to develop a guide for fellow travelers, he described the different species and began to map their ranges as best he could. While working for the American Museum of Natural History in 1926, he continued his work on the guide, using museum records and consulting with Robert Cushman Murphy, and in 1928 he published the first field guide for seabirds, *Birds of the Ocean: A Handbook for Voyagers*, which he later revised and expanded. Using black-and-white photographs and sketches he made from study skins, Alexander compiled a list of seabirds, noting where they lived and how to identify them. He described 4 orders, 15 families, and 290 species, including 13 albatross species, with notes about field marks and how flight actions gave clues to species identity. Seabirds do not fall neatly into one classification, and bird scientists even today do not entirely agree on which birds to call seabirds, but current classification structures align with Alexander's system. Forty pages at the end of *Birds of the Ocean* describe the seas and oceans where seabirds occur. WB concluded that the waters between the Tropic of Capricorn and the Antarctic continent, which he termed the "Southern Oceans," were the home for almost all penguins; most albatrosses, petrels, and cormorants; and half the known gannets, terns, and pelicans.

Collaborative ornithological research was just starting in England when WB was conducting his research. A group working from the University of Oxford sought to establish a permanent center for applied ornithology that would investigate the economic value of various species. WB became director of that fledgling bird group in 1930 and drew support from notable scientists who gathered in London in October 1932 to found the British Trust for Ornithology. The trust sought the university's approval and launched an appeal for funding. The center was named the Edward Grey Institute of Field Ornithology after Viscount Edward Grey of Fallodon, Chancellor of the University of Oxford, because of his strong support for the project. Research projects included bird population variations and the cost of birds' pilfering of crops. WB assembled a first-rate library that now bears his name. He retired in 1945, succeeded by David Lack as director of the institute.

Scientific research and publications and a network of correspondence, comparison, and collaboration accelerated rapidly after World War II. In 1954, based on his own work and Murphy's monumental contributions to seabird ornithology together with those of other seabird specialists, Alexander brought out a new edition of *Birds of the Ocean*. This became a key building block for and contribution to the expanding definition and ornithology of seabirds. By that time, sea watchers, using cliff tops and beaches to observe the passage of birds, were contributing to breakthrough understandings about bird migration. WB, described as "genial, short, round, rubicund,"[21] meshed well with undergraduates at the university, whom he guided on excursions to spot seabirds that he and an equally famous birding brother recorded.

In the 1930s, a New Zealand agricultural instructor in schools, unaware of the work of these earlier scientists, began his own lifelong fascination with albatrosses. In the process, he earned a reputation for being the founder of albatross research. Lancelot Eric Richdale taught mostly in primary or elementary schools for the Otago Education Board in the city of Dunedin on New Zealand's South Island. Specializing in nature study, he visited local schools to introduce children to the area's land and life. One of his pupils remembered him as a quiet and friendly listener who "brought the wonder of nature" into his classes."[22] "Mr. Rich" began to hear from local boys about some albatrosses and penguins that reportedly lived on Taiaroa Head on the Otago Peninsula southeast of Dunedin. His curiosity piqued, Richdale made the first of what became dozens of thirty-mile or so round trips from Dunedin to Taiaroa Head and back to see and then study the tiny group of albatrosses that had apparently selected a patch of the headland's north slope for a nest site. His very first view of the huge, docile, and tenacious Royal Albatrosses, in November 1936, hooked Richdale. He began a lifelong commitment to understanding, protecting, and teaching and writing about his "Royals" of Taiaroa Head.

It is likely that ancestral Maori had hunted albatrosses on Taiaroa Head and on similar headlands prior to European contact. The birds Richdale found on the headland were Northern Royal Albatrosses, which he thought had nested on that section of the South Island east of Dunedin during the previous century, although no direct evidence backed his statement. Some folks had seen albatrosses on the headland in the early 1900s, including six that appeared on the parade ground when troops were garrisoned there during World War I.

Fortunately for the birds, when Richdale arrived, the hundred or so soldiers and dependents garrisoned on Taiaroa Head had departed, leaving only three families to tend the signal light. Reduced human presence and disturbance may have allowed occasional courting and breeding, but Richdale complained that the first nesting pair he observed failed because a local man stole, fried, and ate the egg! The following year, in 1921, the chief signalman removed another one and blew out its contents so he could save it as a keepsake. Until Richdale began round-the-clock guard duties on Taiaroa Head fifteen years later, no breeding pair on the headland was able to hatch a chick. Human intruders stole or smashed all their eggs. In February 1936, when the first young bird managed to struggle free of its shell, it didn't last long. Presumably a stoat killed it.

"I made up my mind to do all possible to prevent a repetition of previous losses,"[23] Richdale declared, and, with assistance from resident signalmen, he kept his word. In September 1937, when more Northern Royal Albatrosses swung down, Richdale set up a camp on Taiaroa Head. "I literally lived along-side the nests and made a special point of being present all day every Saturday and Sunday"[24] when most visitors and picnickers rambled across the headland. Driving nearly five thousand miles between his home in Dunedin and the nest-ing ground that year, Richdale witnessed the birds' entire breeding season. He stood guard over his charges for four months, including their seventy-nine-day incubation and month–long guard stages. Such dedication paid off: two pairs each laid an egg. Although someone throwing a stone at the birds broke one, on February 3, 1938, the chick chipped out of the surviving egg, and 131 days later, as the sun was setting on September 22, 1938, the first recorded Northern Royal Albatross fledgling launched itself from Taiaroa Head, almost a year after its parents had started building their nest. Lurching forward on set wings, the youngster suddenly planed off the headland into a northeast wind. Richdale was elated.

Richdale reported the results of this first breeding season to the Otago branch of the Royal Society of New Zealand and appealed to local townsfolk to help protect the tiny colony. Because most albatrosses nested on remote islands, New Zealanders had limited knowledge of the birds, although from time to time several coastal farmers and fishers had spotted them. What was unique about Richdale's photo lectures to the Royal Society and other groups was his proof of birds actually nesting on the mainland in a spot so accessible to human be-ings. So the branch of the Royal Society appointed him to a committee to raise funds to fence off the area. The committee recruited local lighthouse keepers as unpaid wardens, and in 1938 began a project to trap introduced predators such as ferrets and stoats that had killed at least one chick and continued to threaten every one that hatched.

By World War II, when the military reoccupied the Taiaroa headland and limited public access, Richdale had turned to other seabirds, notably penguins and petrels, and advocated their protection. Eventually, in 1951, with the sup-port of the Otago Harbour Board, the government wildlife service hired a full-time ranger to protect the tiny colony. Stan Sharpe served in this capacity for eighteen years. By then, fifteen chicks had fledged and flown from the headland. Alan Wright took over from 1968 to 1980, and others followed him, until the current head ranger, Lyndon Perriman, joined in 1989. The Harbour Board, which owned much of the headland, negotiated with the government's Wildlife Division for provision of oversight and research. Under the 1953 Wildlife Act, the New Zealand Parliament designated Taiaroa Head a Wildlife Refuge and shared management and protection costs with what eventually became the Otago

Peninsula Trust. In 1964, the New Zealand Wildlife Service (later integrated into the Department of Conservation) officially took charge of bird research and organized scientific studies of the breeding ecology of the Northern Royals on Taiaroa Head. The headland was formally designated as a Nature Reserve, the highest level of protection. The Department of Conservation remains in charge of the colony, allowing the Peninsula Trust to operate a visitor facility that attracts tens of thousands of tourists every year to view "Otago's Royals."[25]

Beginning in the 1930s, Richdale photographed his seabirds—albatrosses, penguins, and petrels—and on his own began banding and monitoring both adults and young. His work with the albatrosses led to experiments and protocols unique to Taiaroa Head. For example, in 1970, Department of Conservation albatross scientist Chris Robertson initiated supplementary feeding of a chick after monitoring concluded the mother was probably dead. Even after its father flew in with food, the young bird continued to beg vigorously, so wardens began to offer the below-weight albatross three to six squid per feeding. The youngster regained and then increased its weight and began jumping and flapping as it exercised in the headland's breezes. On October 10, 1970, the hand-raised albatross sailed off from Taiaroa Head on its first journey over the high seas, returning some five years later to socialize and eventually pair and breed on its nurturing nesting ground.[26]

Personnel at the Taiaroa Head colony continue to feed chicks when necessary with a calibrated diet of marine oils and fish. Rangers have also learned to place dummy eggs in nests of pairs that have lost or broken their eggs to enroll them as foster parents for any orphaned egg or chick. Royals usually accept an abandoned or orphaned chick up to about one month of age and will then continue to care for it as their own. Lessons learned from the seventy or more years of working with Taiaroa Head's Northern Royals are invaluable for recolonization efforts in other places for other albatross species, such as in the Bonin Islands, about 250 miles southeast of Torishima, Japan, where experts are currently working on a ten-year relocation program to introduce Short-tailed Albatross chicks to a former breeding site. Most importantly, New Zealand researchers have generated vital data about the Northern Royal Albatross. In 2006, Ranger Perriman monitored the breeding activities of thirty-one pairs on the headland, and in January 2007, was on hand to welcome the five-hundredth chick hatched there since Lance Richdale had begun his studies. Perriman's knowledge and experience have been especially useful to the relocation project for Short-tails on the Bonin Islands. He and his team continue to conduct careful research and develop additional methods for protecting adults and their fledglings while also protecting their wild natures.[27]

Despite his reserved personality, Richdale slowly began to interact with ornithologists around the world. In 1948, he spent seven weeks on sub-Antarctic Snares Islands south of mainland New Zealand to conduct a pioneering study of the Southern Buller's Albatross. Two years later, with US Fulbright support, he visited colleagues at Cornell University in Ithaca, New York, and the University of California at Berkeley to consult with seabird experts. With funding and access to libraries and papers, Richdale began to publish research notes and monographs, and leading ornithological societies inducted him as an honorary member. For example, in 1952, Richdale published a monograph, *Post-egg Period in Albatrosses*, that gave the first full report on the chick's life from hatching to fledging. Both parents take turns incubating their single egg, with the male frequently taking charge as soon as his mate has laid it. After the chick hatches, the pair continues taking turns for about forty more days, one guarding the hatchling while the other forages at sea. After that, both adults go off to feed and collect food for the chick, which stays alone for longer and longer periods. During this time, the chick inspects its nest area and exercises its wings with bouts of vigorous flapping. It seems to "revel in the wind and rain," noted Richdale.[28] The fledglings begin to move instinctively toward the headland where the optimum sailing wind blows, and when the ready-to-go bird finally catches a forward momentum with its wings at the same moment it jumps up, it planes victoriously out to sea, not to return to the headland for an astonishing four to five years.

Richdale's meticulous notes led him to dispel a persistent misunderstanding that parents abandon their chick, forcing it to fly due to hunger. Quite the contrary, Richdale noted, parents may feed their offspring right up to its last day in the nest. In fact, the chick may become heavier than either of its parents until it slims down from its vigorous preflight jumping and wing flapping and reduced feeds. Parents may even land on the nest site to feed their chick only to find the site empty, and few adults, it appears, actually witness their progeny's departure.

When Richdale received a Nuffield Scholarship to work in the United Kingdom, he established a close bond at the University of Oxford's Edward Grey Institute with bird luminary David Lack, who had succeeded W. B. Alexander as the institute's director after the end of World War II. Lack mentored Richdale and introduced him to other seabird enthusiasts, including Leonard Harrison Matthews. He also commented on drafts of Richdale's publications, which eventually exceeded one hundred books, scientific papers, and popular articles. Richdale also met Lance Tickell, another albatross researcher, who monitored and banded Wandering Albatrosses on South Georgia and, as we shall see, contributed an essential piece to the puzzle regarding the discovery of the albatross's reach and population status.

Knowledge about Richdale's work on the lives of the Northern Royals increased interest among New Zealanders and more and more foreign visitors who wanted to witness the magnificent courting albatrosses as they sailed and slanted in breathtaking aerial display over Taiaroa Head. From its founding in 1967, the Otago Peninsula Trust had lobbied for increased public access to the colony. In February 1972, with Richdale and his wife as guests, the trust opened the first observatory, beside a small ranger hut, and began guided tours for visitors. Some twenty-five thousand visitors viewed courting and nesting albatrosses from this facility during the next ten years. As interest increased and viewers arrived in greater numbers, the Dunedin Rotary Club raised funds to enlarge the albatross center to accommodate the new numbers, and officials opened a new glassed-in viewing area named the Richdale Observatory. Richdale, too ill in Auckland to attend the opening, died only a few weeks after the dedication of the overlook and just a few years before his bird banding effort was to pay off in an astonishing way.

In March 1989, an elegant and more spacious reception center and museum named the Trust Bank Royal Albatross Center (the bank was a big sponsor) opened below the viewing post. Until that time, bird-watchers from Dunedin mostly made the winding trip, but the new center drew in outdoor enthusiasts and nature lovers from all over the world. In its first season alone, more than twenty thousand people drove to the peninsula to glimpse the huge seabirds. It is indeed truly amazing to watch the albatrosses hang almost motionless, like tethered long-winged aerial kites, in a driving wind that may be slanting cold rain or even sleet across the exposed headland, and to realize later, after inching back outside and stumbling toward the car park while trying to stay upright in the gale, that neither bitter cold nor high winds faze the soaring and hovering birds. These denizens of stinging, cold, watery, blustery spaces appear totally at home as they tack and glide in conditions we avoid as harsh and inhospitable.

From the late 1990s, many thousands of visitors were driving to the albatross center annually, which meant ever-increasing appreciation of the Northern Royal Albatrosses as well as of nearby Blue Penguins. However, such an increase in human presence carried unavoidable risks. Rangers noticed that a number of nesting adults were gradually shifting away from the viewing area to sites perhaps less optimal for successful nesting. And chicks of those pairs also tended to move away after returning several years later. Tinting the viewing windows appears to have helped, but there is concern that some pairs may eventually move out of the fenced area. Fire, dry hot weather, and the introduced blowfly threaten both adults and their young chicks. The flies lay maggots on the birds' eggs, and larvae reach inside and parasitize chicks as they are struggling out of the shell. Managers have used crushed mint to disguise the smell of hatching birds and placed rotten meat in flytraps close to nests in order to draw blowflies away from the struggling chicks. Peppermint essence and mechanical incubators to hold the hatching eggs, whose chicks are then placed back in the nests, have reduced fly strike. Ranger Perriman is pleased with the results. However, chicks found dead in the egg on Taiaroa Head and eggshell thinning on the Chatham Islands, the Northern Royals' major breeding site far to the east of Taiaroa Head, raise the possibility of pesticide contamination in the marine food chain.

Lance Richdale launched a public relations coup for the albatross that continues to make Dunedin's Northern Royals a must-do stop on a visit to New Zealand's South Island. More than one hundred and fifty thousand people, more than the population of Dunedin, turn up every year to glimpse the big birds. Taiaroa Head is unique among Southern Hemisphere nest colonies for being easily accessible to the general public, and is, in fact, one of merely a handful in the entire world where admirers can come close to these magnificent birds. Returns from wildlife tourism in Dunedin bring in an estimated NZ$100 million every year. And these albatrosses have captured the imagination not just of scientists and conservationists but also of celebrities, championship sailing teams and round-the-world sailors, royalty, and filmmakers, all of whom continue to raise interest in what is happening to them and their bird cousins in remote sectors of the globe.

In the North Pacific, there are four albatrosses, if you include the Waved Albatross, which nests and forages around the Galápagos Islands a little bit south of the equator. The three other species nest for the most part on four northwest-central Pacific islands: Torishima, Midway, Lisianski, and Laysan (see Map 1). During the twentieth century, scientists like those working in extreme conditions in the Southern Hemisphere have become fascinated with the northern Pacific albatrosses and committed themselves to significant research and conservation efforts. In addition to destruction from human persecution, North Pacific albatrosses have also experienced the loss and transformation of critical nest spaces, notably on Midway Atoll. Midway Atoll consists of five islands totaling over three and one-half square miles. Until recently, Sand Island, which is the largest of the five at just over three square miles, served as a major base and communications link for the US military. The Hawaiian Territorial Government's decision to review the effects of military activities on the Laysan Rail and Laysan Finch, two species released on Midway in hopes of saving them from extinction, lured scientist Harvey Fisher and his wife, Mildred, to the atoll. Working

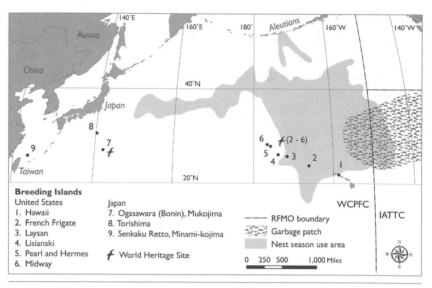

MAP 1. North Pacific Breeding Islands. Major breeding locations are shown for three species: Short-tailed Albatross (mainly 8, some 9, 7); Laysan Albatross (mainly 6 and 3); and Black-footed Albatross (mainly 2–6). Nest season use areas were derived from BirdLife International 2004b, and Regional Fisheries Management Organisation (RFMO) boundaries, from Small 2005.

with a colleague in May 1945, Fisher failed to locate any of the Laysan Rails or Finches, but he quickly grew to admire and respect Midway's albatrosses. The Black-footed Albatross appeared relatively unperturbed by military activities, but he observed that the Laysan Albatross was having problems.

Fisher, his wife, and students from Southern Illinois University, where he chaired the Zoology Department, passed several seasons on Midway, resulting in eighteen publications on the biology and conservation of Laysan and Black-footed Albatrosses.[29] Fisher published the first annotated list of Midway's breeding birds, including factors that were affecting populations. Habitat transformation had hit Laysan Albatrosses hard, he concluded. Bulldozers graded and churned up breeding and loafing areas. Ornamental plants, including shrubs and trees such as sea grape and heliotropes, together with wire fences, limited access and ensnared individuals. Foxholes and gun emplacements acted as death pits: once in them, the large, goofy-footed albatrosses could neither climb nor fly out. Base traffic activity and noise added injury and death, and Fisher believed residents were also killing albatrosses. Some used the birds for target practice. Others cooked their eggs or made shells into souvenirs. Some drivers deliberately ran over the "dumb gooneys," and aircraft landing or taking off hit the long-winged gliders as they competed for airspace over runways. Often, a sleek bird changed into a ball of feathers as it tumbled over and over in a plane's turbulence or hit the aircraft and collapsed in a bloody bundle on the runway. By the 1950s, an average of ten aircraft landed on Midway daily, and researchers figured that at least two of them hit a seabird.

For the next ten years, Fisher, who had devoted more than eighteen months researching ways to minimize conflicts between humans and albatrosses, noted how military officials dealt with the problem of bird collisions—snuffing out lift from steeper-sided dunes, leveling nest areas, paving more and more of Sand Island habitat. He calculated that between 1957 and 1964, bulldozers and workers wielding clubs slaughtered at least sixty thousand adult Laysans. Another three thousand to seven thousand birds had succumbed to other types of mishaps. Together, these losses had cut Midway's overall breeding population in half. As Fisher saw it, this destruction was not necessary. He made the radical suggestion to military planners that they consider organizing human activities around the birds, which would mean, among other changes, flying aircraft at night or during times of lowest bird movement.

Not surprisingly, authorities dismissed his suggestions. Fisher's earliest survey had spelled out a problem, but albatrosses, not humans, were expected to adapt. And many had adapted, by nesting under ironwood trees planted by the cable company or in residential gardens and yards. They were also adjusting to dogs that sniffed and growled at them and to people who handled and played with the chicks. Children loved the wobbly, fuzzy chicks and collected them in carts, tugging them across lawns and down streets like friendly pets. Fisher's wife, Mildred, came upon a bunch of three- and four-year-olds hauling along thirteen mop-shaped gooney chicks they had pried, one after another, from beneath their brooding parents. They were giving them a joy ride.

Fisher noted also that invasive species were killing these and other ground-nesting seabirds. He investigated house rats, which had scampered ashore in 1943 when traffic to the island had grown plentiful. Unlike other islands where rodents actually gnawed live tissue of nesting birds, Midway's house rats and equally populous house mice made do with smaller, burrow-dwelling Bonin Petrels and Bullwer's Petrels. Rats had eaten the last of the Laysan Rails the year Fisher arrived to check on them. Rats also lived on garbage and scavenged bird carcasses. People heaped together dead albatrosses, which drew swarms of flies and gave the hungry rats easy pickings, much to families' disgust. Finally, officials began to set out poisons, but it took fifty years to eradicate the noxious rodents.

Carl Safina, founder of the Blue Ocean Institute, spent a week on Midway in the late 1990s. Safina talked to US Department of Agriculture employee Jim Murphy, who claimed to have finally killed off the sneaky, destructive rats. Murphy used sweat, stealth, and chemicals, he said. First, he cleaned up after the big birds had left the atoll, then set out poison baits on a fifty-meter grid, placing traps between them. "A little bit of trapping," noted Murphy, took "all the big boys out" and shattered the Midway rat society. Four months of continuous trapping killed off 99.9 percent of them, and two years more of baited stations poisoned the remainder.[30]

But back in the 1950s, Laysan Albatrosses on Eastern Island had been introduced to a new kind of death and dying, a forest of towers and wires that made up a PACSCATTER relay station. This mishmash of metal wires stretched and slung between large steel girders proved a death trap for passing Laysan Albatrosses, and maiming and killing began as soon as the structure was in place. Dismayed by how many seabirds flew into this structure, and cursing his inability to make authorities modify it, Fisher lamented that "the world may lose one-sixth of its [population of] Laysan Albatrosses because of one small forest of antennas at Midway Atoll in the North Pacific."[31] When Fisher studied the problem a decade later, he was sickened by what PACSCATTER had done to the colony of thirty thousand Laysans on Eastern Island. Storms took them out by the hundreds, and overcast weather made the apparatus more lethal. Fisher had recommended the use of plastic streamers on the towers and wires to make the obstacles more conspicuous, only to be told that such streamers might impair the structure's integrity. He commented that military engineers apparently thought a gigantic bird hitting a wire or pylon at fifty miles per hour posed no structural problem. When he revisited the antenna issue in 1970, he was happy to discover that space relay satellites had made the bird-killing forest redundant, and the PACSCATTER facility on Eastern Island was to be demolished. Plastic explosives did the job. But even though workers shifted albatross chicks from under the antennas, the blast and falling debris killed thousands of Sooty Terns nesting under the structure's shadows. The pile of twisted metal was dumped into the sea.

In June 1997, when Midway Atoll became a National Wildlife Refuge, workers tore down buildings and ripped up roads and runway tarmac. For five years thereafter, an ecotour operator beckoned bird-watchers and scuba divers to the atoll and its confiding animals, including fifteen species of seabirds breeding on its two-square-mile surface. Reefs glittered with hundreds of variously colored fish and other marine creatures. Some visitors began visiting Midway on an annual basis. After incurring losses, however, the firm pulled out abruptly in

ABOVE: Laysan Albatrosses have flourished on Midway Atoll since the US military left; however, longlines, plastic objects fed to chicks from the "garbage patch" northwest of the atoll, and ingested lead paint take their toll. Here Laysan Albatross nests are being counted. Courtesy US Fish and Wildlife Service. BELOW: An example of a dead albatross chick compacted with floating objects taken as food. Photo © Chris Jordan.

January 2002. Recently, the Oceanic Society has begun to offer weeklong stays on Midway from March through May, with flights from Honolulu, limited to groups of fifteen visitors at a time. A cadre of scientists and volunteers working on Midway continue the Fishers' work by monitoring biological processes, events, and trends. Currently, Midway Island is the quietest it's been in three generations. That's how the albatrosses found it and presumably that's how the current generation of birds gathering on Sand Island and Eastern Island prefers it, free of the confusion, dangers, and mistreatment their immediate ancestors experienced. However, the population of Laysan Albatrosses is again unstable, and experts are concerned.

On Torishima, an island about 360 miles south of Tokyo with an active volcano at its center, the dire fate of the Short-tailed Albatross has captured the full commitment of biologist Hiroshi Hasegawa. The Short-tailed or Steller's Albatross was named after the intrepid German collector-naturalist Georg Wilhelm Steller, who traveled across Russia as far as the northeast Pacific coast in search of the eastern passage to North America. Along the way, he collected specimens of Short-tailed Albatross and of other water birds off Kamchatka. Steller died in Russia, but his bird skins reached Saint Petersburg, where a colleague, Peter Pallas, described them and named the lovely white-headed albatross after Steller. When Steller visited, this Northern Hemisphere seabird cruised the western Pacific Ocean, including Alaskan waters, breeding on small islands around Japan. It also commonly soared northward into the Aleutian chain, and records from Native American coastal middens place Short-tails as far south as California. Today, however, because of prolonged and intense persecution on Torishima, the Short-tailed Albatross is a rare but recovering species. In 1950, the chief meteorologist in charge of Torishima's new weather station counted several nests along the south coast, and three years later photographed thirteen adult-plumaged Short-tails in their ancestral nest area. But high winds and heavy rain occur frequently in those subtropical latitudes, and a typhoon blasted the island in 1958, killing adult birds just as recovery efforts were getting started.

Over the following decade, scientists from the Yamashina Institute for Ornithology, along with Tokyo metropolitan officials, worked to protect the surviving population. When weather station employees quit Torishima in 1965 after earth tremors posed an imminent threat of an eruption, they noted that albatrosses had begun to successfully reoccupy an ash-strewn slope and were tending eleven well-grown young.

With support from the International Council for Bird Preservation and the New York Zoological Society, in late April 1973, albatross researcher Lance

Tickell put ashore from the Royal Navy vessel HMS *Brighton* and camped for a week in the abandoned meteorological station. He and his colleague, the chief of the Japanese bird-banding center, counted twenty-four chicks among fifty-two pairs, a world population of fewer than two hundred individuals. Tickell was concerned with the threat from rats on the island and also noticed a fat ginger and white cat skulking in the lava field filled with storm-petrel nests, suggesting "that it lived by catching petrels." He expressed concern for the distant albatrosses, since "even one cat is a potential menace."[32]

Tickell also met Hiroshi Hasegawa, a young biologist at Toho University in Chiba, Japan, and spoke to him about his brief time on Torishima. From childhood, Hasegawa had been fascinated by rare species, especially the Crested Ibis and the sadly diminished albatross, the latter of which he selected to study, directing his professional life to conserving the beautifully plumaged seabird on its major nest island.

One of Hasegawa's first projects was to stabilize the friable ash slope in the nest area perched on the island's south end, where some eggs had the misfortune of rolling downslope into the sea. He planted sod-forming *Miscanthus* grasses only to have a mudflow destroy both eggs and chicks in 1987. Further landslips happened in the early 1990s and lowered breeding success. In 1995, a storm struck just before the nesting time and scoured the vegetation Hasegawa had planted.

Upslope of the preferred nesting site, Hasegawa built a rock dam to avert flooding, channeled rainwater, and dug fifty or so terraces to stabilize the nesting slope, filling wire berms with rocks to protect against mudslides. The big birds began fledging more chicks. Earthmoving machinery failed to entirely stop erosion on the old site, but management there has also nudged fledging success upward. Hasegawa also implemented a plan to lure his birds from their unstable site to a safer place about a mile away by setting up twenty or so life-sized albatross figures and playing tapes of Short-tail calls. Some adult and immature birds responded, and in 1995, a single pair moved in close to the old meteorological station where observers had transplanted grasses forty years earlier. By 2003, the pair on this decoy site had fledged six chicks, and in December 2004, three additional pairs moved in and bred successfully. All four pairs each raised a chick in 2005, and two years later that number had risen to sixteen.[33]

In May 2002, when Hasegawa made his eightieth visit to Torishima, he estimated that 1,415 birds occupied the main colony and an additional 260 sat on the uninhabited but internationally disputed Senkaku Islands, which are about one hundred miles northeast of Taiwan. Five years later, the total population of this

41

once critically endangered species had jumped to 500 birds on Torishima and almost 200 on the Senkakus, enabling Hasegawa to confirm that this Special National Monument bird, designated as such by the government in 1962, was firmly on the road to recovery. Numbers have continued to rise on Torishima, and the overall population is now closing in on 2,500 birds. Numbers have increased on Senkaku as well.

This is a wonderful increase from 1954, when seven eggs were the first ones reported in modern times. The threats of torrential rains and volcanic eruptions are ever present, but Hasegawa is no longer alarmed by these natural phenomena. Young birds stay at sea for at least four years before coming ashore to visit their birthplace, and there is always a pool of nonbreeding adults roaming the North Pacific Ocean, amounting to about 20–25 percent of the total adult population. Hasegawa surmises that these birds will wait until conditions are suitable before landing to nest on the ashy slopes.

Short-tailed Albatrosses are being transported from Torishima to Mukojima on the Bonin Islands, a previous nest site. Dummy birds and recorded calls assist these hand-fed chicks to familiarize themselves with the new location and, after several years, return and begin to nest on the island. Courtesy Lyndon Perriman.

Hasegawa has conducted Buddhist ceremonies over the human settlement entombed by a 1902 volcanic eruption on Torishima. In 1987, some one hundred years after workers arrived to slaughter birds for their feathers, he placed albatross bones on a stone near two memorial cairns, or stupas, lit candles, burned incense as an offering to the ancestral albatrosses, poured seawater for bird spirits to drink, and set out favorite foods.[34]

In the first decade of the twenty-first century, breeding has been confirmed on the Bonin Islands, also known as Ogasawara Gunto, about 180 miles south of Torishima, which are a historic breeding archipelago that albatrosses increasingly visit. Discussion has centered on reintroducing birds to northwest Mukojima of the Bonin group by placing decoys to bring down passing birds and relocating as many as ten young chicks a month or so old to a new site. In earlier experiments in removing albatross chicks from Midway to other sites, many birds that survived returned to their original birth sites on Midway, not to the islands where experts had transported them. With this in mind, one idea has been to relocate Torishima-born Black-footed Albatross chicks to Bonin in order to develop effective protocols before placing any live Short-tails there. In another experiment, in early March 2006, a charter plane landed with ten Laysan Albatross chicks on Kauai, Hawaii, to be hand-reared on the Kilauea Point Refuge, where that species already nests, as a prelude to possibly repeating the experiment with Short-tail counterparts on a safer site than Torishima. Japanese researchers teamed up with Hawaiian wildlife biologists to feed and tend the young birds through to fledging, but infections dogged the birds and not many fledged.

However, in May 2008, ten Short-tails, transported from Torishima and hand-reared on Bonin, where decoys and taped calls were used to help the chicks imprint on their species, fledged from Bonin's uninhabited Mukojima Island. Two years later, a total of forty transferred chicks had fledged there, and tagging suggested many had moved toward the Aleutians and Bering Sea, usual foraging areas for the species. Workers are confident that at least some Short-tails will prove pioneers and begin to nest once more on the Bonin Islands. The government has submitted the island group for World Heritage designation.[35]

CHAPTER 4

Connections

IN THE LATE 1930S, Lance Richdale began banding albatrosses on Taiaroa Head, setting in motion a method for identifying and tracking individuals. Little did he dream where his work would eventually lead. Robert C. Murphy's observation of Wandering Albatrosses on the high seas notwithstanding, most people assumed the birds foraged near their nesting colonies. Richdale's banding of albatrosses initiated a practice that soon yielded surprising fruit in expanding and mapping globally our comprehension of the reach in space and time of these great birds. With new technology, it is still going on.

Richdale's banding also yielded the story of Grandma and her chick Button, narrated by Sir David Attenborough in his 1990 television documentary about Taiaroa Head's colony, *Grandma: the Oldest Albatross*. Grandma, then known as Main Path Female and one of the first birds Richdale banded, had paired with a banded male. The couple became known as the Favorites and remained together eighteen years, until 1955. Thereafter Main Path Female mated serially with three additional males until she disappeared at sea while foraging for Button in June 1989. At that time, Grandma was at least sixty-two years old and had outlived Richdale, her human guardian. Calculations support the claim that Grandma was the longest-living albatross ever banded. Indeed, she is among the oldest birds recorded in the world. Button reappeared at the nesting colony on Taiaroa Head as recently as 2010, a living reminder of his famous mother. But the practice of banding albatrosses that Richdale started yielded a much larger and more global story than this.[1]

Because of international interest in the Antarctic in the 1950s and the need for a more scientific basis for planning and decision making, governments began sponsoring scientific expeditions to remote Southern Ocean archipelagoes to take stock of plants and animals in what had in many instances been seal butchering sites. In so doing, they began to uncover some of the mysteries surrounding the lives of seabirds, including albatrosses. South Georgia played a supporting role in the 1957 International Geophysical Year, which focused on the globe's coldest localities. Scientists from many nations explored seismic, gravitational, and glaciological phenomena in the Antarctic using the most advanced new technology. In 1958, Bristol-based zoologist Lance Tickell sailed to South Georgia to begin studying albatrosses under Johns Hopkins University sponsorship, complementing Geophysical Year research.

Tickell's lifelong passion for albatrosses resulted in his 448–page *Albatrosses*, published by Yale University Press in 2000, the most thorough compendium of knowledge about albatrosses that exists to date. *Albatrosses* is a historical and biological survey drawn from years of Tickell's own fieldwork and a full grasp of the work of others. Tickell depended on and consistently acknowledges the contributions of others to his work: over two hundred fellow scientists, albatross specialists, marine area and resource experts, fishers and their associations, research centers, commercial operations, publications, and funding sources that form an intricate and enduring global network that has supported and furthered our understanding and protection of the albatross, a network Tickell himself helped establish and in which he has participated for more than forty years. As a result of Tickell's commitment and that of his many generous colleagues, research and publication of findings have accelerated rapidly over the past twenty years.[2]

Tickell led expeditions to South Georgia for six years after his first austral summer there in 1958, working primarily in the large Wandering Albatross colony on Bird Island, a three-mile-long and half-mile-wide stretch of rock, lichen, and tussock grass on South Georgia's northwest tip. In late November, he and his colleagues erected a field hut and began to band all the well-grown Wandering Albatross chicks they could find, eventually fitting metal leg bands on close to seven thousand birds.

On his first banding project, Tickell made an astonishing discovery. In December 1958, as he walked through a colony of about five thousand nesting birds, he was astonished to see a Wanderer that was already wearing a leg band. He was even more amazed to discover that the band had been put on in Australian waters months earlier. Using the information on the band, he was able to contact Doug Gibson, an Australian seabird enthusiast and albatross specialist, who told him that the bird had been banded the preceding August off the coast of New South Wales, south of Sydney, Australia. By the shortest sea route, the bird had flown more than seven thousand miles in four months from Australia to Bird Island in the South Georgia archipelago.

Making the Australian connection was possible because of the work of amateur ornithologist J. Douglas Gibson, who worked in New South Wales's Port Kembla steel plant. Together with Alan Sefton, who beach-walked for stranded birds, and other fellow birders such as Henry Battam, Gibson had taken to watching and photographing albatrosses along coastal New South Wales. Beginning in June 1951, observers had made regular visits to the 150-foot cliffs overlooking Sydney's Malabar sewage outfall zone where albatrosses were turning up annually from late April through mid-November. Malabar drained Homebush Abattoirs, tanneries, food-processing plants, and other rendering facilities between Sydney and Botany Bay and provided abundant waste for hungry seabirds. From their cliff perches, the men caught startling views of the giant albatrosses wheeling, dipping, swimming, and squabbling over the floating detritus. They reported an "'albatross density' apparently not equaled elsewhere in the world,"[3] with as many as four hundred Wandering Albatrosses circling and swimming in the outflow as it curled from the cliff face and moved south in the longshore drift.

South of Sydney, other bird-watchers began surveying a reef that paralleled the shore off Thirroul, where D. H. Lawrence lived while writing *Kangaroo* and where Gibson resided. In 1958, the watchers off Thirroul, some forty miles south of Sydney in the northern sector of the city of Wollongong, joined with the Sydney-based albatross seekers to form the New South Wales Albatross Study Group as part of a new Australian bird-banding program. In 1994, the Albatross Study Group was renamed Southern Ocean Seabirds Study Association, or SOSSA. Its members continue to band albatrosses today.

The birds' habit of riding close to the sheer cliffs around Malabar and milling over Thirroul's close-in reef dispelled all assumptions heretofore that albatrosses were lonely wanderers who sought out vast, isolated spaces, thereby distancing themselves from human observation. In August 1954, Gibson and Sefton observed albatrosses, mostly Wanderers, along a three-mile strip off Thirroul. Working with a boat master from nearby Bellambi, also within the city of Wollongong, the men maneuvered a small fishing vessel to watch the huge seabirds tearing into the large cuttlefish that had gathered to spawn and then die. Albatrosses and dolphins were taking both live and moribund cephalopods, many of which floated near the surface. The huge birds pounced eagerly on carcasses that might weigh in at thirty pounds and dipped for snacks spun up by the speedy dolphins. Close and repeated observations of these birds led Gibson to publish an important article about plumage changes and size differences in Wandering Albatrosses and in a subspecies that some regard as a full species, now aptly named Gibson's Albatross.

Gibson recalled that in August 1955, one of the group, Doctor A. M. Gwynn, suggested banding the birds off Bellambi in order to identify individual albatrosses and monitor them as they returned year after year, detailing migration as best they could from one fixed spot. They knew that systematic banding and monitoring could reveal patterns of long-distance movement and details about the albatross's life history, but they had no idea how far "long-distance" would come to mean. First they faced the not negligible challenge of grabbing the huge, thrashing seabirds and fastening a band over one of their legs.[4] They tried floating a wooden board toward the birds that was fitted with a metal spring-hinged triangle baited with cuttlefish. Theoretically, the triangle would snap and clamp the bill of any albatross that pecked at the bait. But the mouse-trap approach didn't work, so they quickly switched to hand-thrown nets. One man steered the small boat downwind toward a swimming bird, and as he cautiously eased to within twenty feet or so, the catcher in the bow stood up and hurled a hoop net at his target. This worked well. Once a bird's head was tangled, the banders hauled it aboard—"haul" being the operative word—and quickly grabbed the menacing beak to stop its owner from slashing an arm or spewing oily stomach contents all over the boat. Fortunately, after a few minutes, the albatross usually relaxed and became easier to handle. The birds' docility around humans has led to many problems, but in this case, it facilitated what would lead to a systematic and scientific study of dispersion and flight patterns as well as, eventually, the bird's population status. With practice, netters learned to catch as many as fifteen albatrosses an hour and fit them with bands with a serial number and the words "Write-Wildlife CSIRO Canberra-Australia."[5] The men also marked the head of every banded bird with a gentian violet dye to indicate it had been banded. Greater care was needed for the offal stream near Malabar, where a two-man team bobbed in the polluted outflow as they steered toward the clusters of flapping, feeding birds in a smaller, less stable dinghy than the one birders sailed in off Thirroul. They moved very, very gingerly to avoid swamping the boat, first when capturing the big birds and then when fixing a metal band on a leg of every captive. From 1956 to 1958, the albatross seamen captured and banded 197 Wandering Albatrosses, 125 of them at Bellambi.[6]

It was one of these Bellambi Wanderers that connected the dots between the waters off Sydney and Bird Island in South Georgia, seven thousand miles away on the far side of the Southern Ocean. Mirroring this exciting discovery just over six months later in July 1959, Australian enthusiasts discovered a reciprocal passage of birds at sea off Wollongong when they picked up a Wanderer off Sydney sporting a South Georgia band that dated from Tickell's work the previous December. These two birds literally opened the world of albatrosses to scientists and enthusiasts. Both sets of workers realized that Wanderers were flying thousands of miles on a regular basis between nest sites, in this instance from Bird Island in the South Atlantic Ocean to a nonbreeding area off New South Wales in the South Pacific Ocean. This challenged the image and concept that these birds wandered randomly as vagabonds. At least some individuals were not wandering but in fact commuting from their home colonies to a specific and memorized feeding site a long way away. Once they reached Australian waters, for example, observers noted that the big albatrosses passed weeks, even months, foraging over a relatively small stretch of water before clearing off. Clearly these birds knew where the food they wanted was and how to navigate across huge stretches of seemingly monotonous ocean in order to arrive at the same places every winter.[7]

Bird Island, South Georgia. The present British Antarctic Survey base continues the long-term study of albatross population trends begun by Tickell and later colleagues. Courtesy of Chris Gilbert, British Antarctic Survey.

Tickell recalled that after the initial identification of his Bird Island albatrosses with those feeding off Australia, no fewer than fourteen additional individuals from Bird Island turned up in the same Australian waters in the following six years. During that time, workers on South Georgia identified another fifty Wanderers banded in Australia. Under US National Science Foundation sponsorship, Tickell made a second trip to Bird Island in 1960 to band several thousand more seabirds, mostly albatrosses, and to take blood and food samples to study. His group also painted pink dye on seventy-five Wandering Albatrosses in order to track them, and began receiving reports of pink albatrosses as far as the Falklands and South Shetlands. Tickell made another lengthy visit in December 1962 to overwinter on Bird Island to study Wandering and Grey-headed chicks, the product of their parents' biennial breeding. He continued observing and banding birds on Bird Island until April 1964. He filmed and made sound recordings of the area seabirds, analyzed their stomach contents for dietary purposes, and tested nest recognition by adult birds.

In all, over six summers and one winter, Tickell and colleagues on South Georgia banded about fifty-three thousand individual seabirds of seven species. He found that the connector albatross that had flown from Australia to nest in December of 1958 had remained in the Bird Island colony. It was absent in the 1962–1963 season, but Tickell recorded it again on Bird Island during the 1963–1964 nesting season and learned that Australian colleagues had recorded it in the usual Australian haunts prior to its return again to South Georgia.[8]

When Tickell concluded his work on Bird Island, Gibson and others had banded more than seventeen hundred Wanderers along the coast of New South Wales and tallied sixty-one recoveries, mostly from Bird Island. They figured that some adults made the round trip once every other year, and that almost half took less than six months to make a one-way flight. Because of banding efforts elsewhere, the albatross group was also able to identify albatrosses in their waters from other islands such as Kerguelen and Crozet in the Indian Ocean and off New Zealand. Some of these non–South Georgia birds also wintered in Australian waters year after year, and one Wanderer from France's Crozet made seven seasonal appearances.[9] Calling up an old record from the 1840s, Tickell confirmed a tagged bird's travels of 134 miles per day over a three-week period, or nearly three thousand miles. He surmised that it was in fact possible for an albatross to make the long sky trek from Australia to South Georgia or vice versa in merely two months, and that nonbreeding Wanderers could "circumnavigate the Southern Hemisphere several times in a year."[10] Eventually, Tickell made his own albatross-type journey to Malabar and Bellambi, but unfortunately not in

49

time to meet Doug Gibson. He was, however, able to visit with Henry Battam, Gibson's coworker. By that time, albatross numbers in the Australian waters had fallen dramatically because of the closure of the Malabar stream, and they have remained low compared with the heady days of banding by our cockleshell heroes. Lindsey Smith, who has guided pelagic trips from Wollongong to study and band birds, reports sighting many fewer albatrosses than in the late 1970s and doubts that all species will survive. He believes the first to go will be the Wanderers on Bird Island—a sad end to such exciting discoveries.[11]

After the flurry of observations and bandings from the late 1950s through the mid-1960s, the tundra-looking island upon which the giant Wanderers had their nest colonies remained unobserved until British Antarctic Survey personnel passed two seasons there in the early 1970s. Living up to the name Captain James Cook provided in 1775, and because of Tickell's careful long-term work, Bird Island has been a critical study site for measuring the population trends and status of Wandering Albatrosses. British Antarctic Survey personnel began to count nesting birds there regularly after 1971 and established all-year occupancy after 1982. As early as the 1960s, scientist Pierre Jouventin, joined by Henri Weimerskirch, J. C. Stahl, J. L. Mougin, and others, began studying seabirds of the French possessions and Adélie Land on Antarctica for the Institut des Sciences de L'Evolution, Laboratoire de Socioecologie in Montpelier, and the Museum National d'Histoire Naturelle in Paris. And by the 1970s, South African, Australian, and New Zealand bird scientists had begun systematic research on albatrosses in the Southern Hemisphere, noting with some relief that populations appeared to be recovering from the era of destruction at the hands of sealers and whalers.

CHAPTER 5

Home

OF THE ALBATROSS SPECIES that appear in the Southern Hemisphere, only one, the Waved Albatross, which nests mostly in the Galápagos Islands, does not breed on islands in or close to the Southern Ocean (see Map 1). Only one of the three albatrosses from the Northern Hemisphere, the Laysan Albatross, has been observed in the Southern Hemisphere, but rarely. Thus, the relatively unknown southern reaches of the Atlantic, Indian, and Pacific Oceans are the waters that concern us the most in telling this story. Having a mental picture of the Southern Ocean makes it easier to place most of the albatross species that are currently at such great risk. It also makes it easier for us to visualize, later in the story, the overlap of albatross nesting and foraging habitat with key commercial Bluefin Tuna and Patagonian Toothfish fisheries and fishery management jurisdictions. It also assists in grasping the geography and extent of illegal commercial fishing in that part of the world. Most important, it makes obvious the inextricably linked nature of albatross and fish conservation as well as the necessarily international scope of such conservation. An additional pleasure is that it allows us to see the world through the eyes of the albatross.

The Antarctic Treaty, which came into force in 1961, defined the Southern Ocean as that part of the Atlantic, Indian, and Pacific Oceans between 60°S latitude and the Antarctic landmass. In 2000, the International Hydrographic Organization officially named this southern terminus for the three oceans the Southern Ocean. This circumpolar marine body of almost eight million square miles is the world's fourth largest ocean, about twice the size of the United States or of the Arctic Ocean, which is the world's smallest ocean. For the most part, the Southern Ocean is twenty-five hundred to three thousand feet deep with shallow, crimped continental edges. However, underwater abysses and trenches are as deep or deeper than North America's Rocky Mountains are high. Seamounts, underwater pinnacles and plateaus, which rise to within six hundred feet of the surface east and south of New Zealand and form the Patagonian Shelf off southeast South America, also vary from one to two thousand feet deep over the fourteen-hundred-mile-long Kerguelen Plateau in the Indian Ocean. The Kerguelen Plateau is the world's longest underwater mountain, one of many that interrupt the corrugated abyssal ocean floor. These subsurface ridges and slopes, together with the taller, slender seamounts, block the free passage of marine waters and create vertical mixing and strong cross-flows that provide especially rich habitat for shallow through deepwater marine life. It is in these mixing waters that fishers and albatrosses encounter each other.

Phytoplankton, floating microscopic plants, are the essential components in the food chain. Zooplankton graze on them and in turn become food for the essential krill on which feed long-lived and cold-water-adapted toothfish and icefish, mammals such as whales, and birds including albatrosses. Krill, derived from the Norwegian *kril* for "young fish," are two-inch-long shrimplike crustaceans that swarm in massive but patchy agglomerations. They are the keystone species found closest to Antarctica in the Southern Ocean, providing several hundred million tons of staple food for vertebrates, including our seabirds.

Antarctica, the White Continent, is the least human populated and most remote, unspoiled, and southerly of the globe's seven continents. It comprises 10 percent of the earth's land area, the size of the United States and Europe combined, or about 5.5 million square miles. Inland, frigid conditions and an ice mantle that holds 75 percent of the world's fresh water make life inhospitable for land animals and most plants except for simple lichens, mosses, and liverworts. Productivity varies in the polar extremities of the Southern Ocean. In winter, sea ice covers almost 6.5 million square miles of Antarctica's ocean edge and closes off surface foods out to twelve hundred miles from the shore, forcing albatrosses and other seabirds to move farther north to forage. In November, the sea ice breaks up and opens areas for krill-eating birds, leaving a fringe of sea ice less than one million square miles along the coasts. With longer days and more sunlight, blooms of plankton spread over the newly exposed water and attract several species of krill that color the water pink as far as one hundred feet down as they graze on tiny organisms.

North of the widely variable Antarctic ice shelf lies the Antarctic maritime zone, the dynamic northern boundary of which is a stretch of the Southern Ocean that encircles Antarctica at roughly 55° to 58°S latitude, deviating in places through upwelling, crosscurrents, and gyres, bands of water that spiral both north and south in huge curlicues. This line of shearing and mixing is called the Antarctic Convergence or, more technically, the South Polar Frontal Zone, where the cold waters flowing north off the Antarctic landmass converge with and sink under warmer flows from more temperate latitudes. The Frontal Zone creates a biological barrier around the globe, bounding what is largely a closed Southern Ocean ecosystem. This boundary defines the northern limit for krill, on which so many seabirds and marine mammals depend. Beyond the Frontal Zone, squid predominate in seabird diets.

For purposes of scientific investigation and environmental management, this Antarctic Convergence forms a biological boundary for the Southern Ocean more specific than the official static Southern Ocean limit of 60°S defined by the Antarctic Treaty. But it also serves our purposes to look farther north to encompass the nesting and foraging locations of all the Southern Ocean albatrosses. The nest islands of some of these birds, together with those of petrel cousins and penguins, spread from close to the Frontal Zone to north of 60°S and are under the jurisdiction of six nations: Australia, New Zealand, the United Kingdom, South Africa, France, and Chile (Map 2). This pushes our study area beyond the Frontal Zone by almost twenty degrees, or 1,200 nautical miles, to include breeding islands from Tristan da Cunha and Gough Island at 37° and 40°S latitude respectively all the way down to Macquarie Island at 54°S latitude, 600 miles southwest of New Zealand and 932 miles southeast of Hobart, Tasmania, and even farther south to Chile's Islas Diego Ramírez at almost 57°S, southwest of Cape Horn. Albatrosses home in to court and breed on these tiny pieces of usually volcanic crust that tip above the ocean surface in this latitudinal band of southern waters, and they forage for food both north and south of the Frontal Zone in the adjacent seas, in some cases almost as far north as the equator. The eight million square miles of Southern Ocean south of the Frontal Zone fall under the jurisdiction of the Convention for the Conservation of Antarctic Marine Living Resources (CCAMLR), whose northern boundary is officially and unfortunately limited to the Antarctic Convergence. Birds also inhabit, without being restricted to, what people officially term the exclusive economic zones around the sub-Antarctic island possessions of France, South Africa, the United Kingdom, Australia, and New Zealand, all of which today generally comply with CCAMLR regulations with respect to conservation of marine resources.[1]

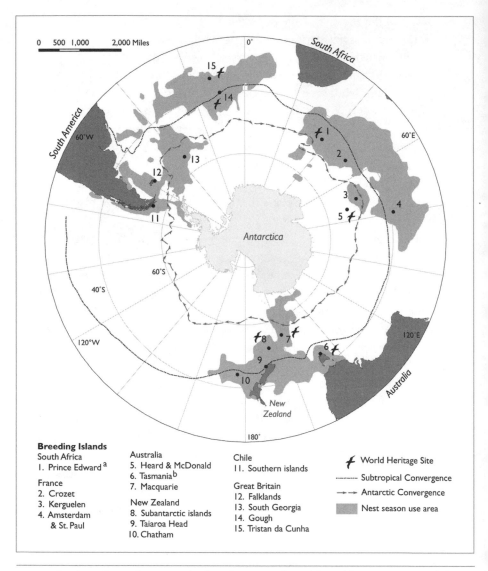

0   500 1,000   2,000 Miles

0°

South Africa

South America

60°W

60°E

15

14

1

2

13

3

12

4

5

11

Antarctica

60°S

40°S

120°W

120°E

8   7

6

9

10

New
Zealand

Australia

180°

Breeding Islands

South Africa
1. Prince Edward [a]

France
2. Crozet
3. Kerguelen
4. Amsterdam
   & St. Paul

Australia
5. Heard & McDonald
6. Tasmania [b]
7. Macquarie

New Zealand
8. Subantarctic islands
9. Taiaroa Head
10. Chatham

Chile
11. Southern islands

Great Britain
12. Falklands
13. South Georgia
14. Gough
15. Tristan da Cunha

𝄐 World Heritage Site

---------- Subtropical Convergence

→→ Antarctic Convergence

Nest season use area

MAP 2. Southern Hemisphere Nest Islands and Marine Habitat. (a) World Heritage proposed; (b) only two Tasmanian nest islands have World Heritage status. Nest season use adapted from BirdLife International 2004b, and ocean convergences, from Harris and Orsi 2001.

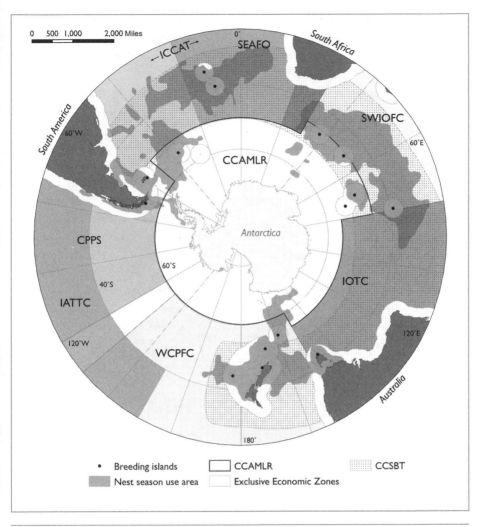

MAP 3. Regional Fishery Bodies and Albatross Nest Areas. RFMOs, shown in gray, widely overlap in some cases. Nest season areas were derived from BirdLife International 2004b.

Most breeding birds that inhabit waters north of the Frontal Zone and CCAMLR jurisdiction come under the regulatory authority of tuna fishery management organizations, especially the Convention for the Conservation of Southern Blue-fin Tuna (Map 3).

As we've said, in the Frontal Zone, cold, brine-laden outflows released below the sea ice from the Antarctic edge collide, merge with, and then sink under slightly warmer saline currents. As the colder waters sink to about three thousand feet, they join bottom waters to form a dense, briny, frigid current that creeps northward at about ten million cubic feet per second. Meanwhile, the southbound sub-Antarctic surface waters mingle with lighter, less brackish waters to remain on the surface, heading generally in an easterly direction with backward-turning ebbs, flowing, for example, between Kerguelen and the paired Australian islands Heard and McDonald (HIMI), and closing in on Australia's Macquarie Island from the southwest (see Map 2).

North of the Frontal Zone, seawater temperatures can change quite suddenly. In the age of sail, mariners noted a shift as they passed from the strong icy winds and rolling seascape of the Frontal Zone into the more genial air north of the convergence, usually accompanied by a gentle rain, like England in springtime, some recalled. These are the 50°S latitudes of the Falklands, the Prince Edward Islands, Crozet, and New Zealand's Campbell and Auckland archipelagoes, all vital albatross nesting places. The colder waters continue moving northward, circulating below the warmer waters, which may reach 58° Fahrenheit, creating what hydrologists call the Subtropical Convergence, a secondary boundary of major mixing where sub-Antarctic flows merge with waters coming south from the tropics. This is not an easy place to sail, especially in what are known as the Roaring Forties. Low-pressure cells of air spin from the west and add extra turbulence to this mixing and nutrient exchange. But albatrosses love to patrol these waters.

In addition to the energy of these two convergences, the earth turning on its axis pushes ocean waters in what is called the Antarctic Circumpolar Current, the mightiest of all ocean currents, carrying one hundred times more water than all the earth's freshwater rivers combined. In some areas, it carries a column of water thirteen thousand feet deep, in flow spans that may spread out as wide as 1,400 miles or more. This Circumpolar Current spins around the globe between 45° and 65°S along a track more than 12,400 miles long, unobstructed by any landmasses, though it narrows and accelerates when it blasts through the Drake Passage between the tip of South America and the Antarctic Peninsula. The Current carries an estimated four million cubic feet of water every second perpetually eastward, one thousand times the rate of the Amazon River's flow. A bottle dropped into the Antarctic Circumpolar Current bobs eastwardly around the planet for about eight months before reaching its starting point again. Albatrosses know how to work this surge and race.

High winds in what some call the Furious Fifties and Screaming Sixties, where albatrosses appear to be in their element, further heap and tumble the upper layers of this marine swirl, creating what is known as the West Wind Drift. This giant spiraling ocean current pulses, meanders, and loops like a hose around Antarctica. Although albatrosses prefer the eddies and curlicues in the energy-rich flows within the Circumpolar Current, they also tack far south and north of it in search of food. The Wandering Albatross, for example, may fly into ice-filled waters off Antarctica and then veer north into and beyond the Subtropical Convergence, cruising upwelling along the west coast of Africa or the east and south coasts of South America.

The Circumpolar Current's massive surface flow and subsurface exchanges transfer heat, salt, oxygen, carbon dioxide, and minerals to ocean basins all over the planet. The Current is a lifeline to other oceans and influences global climate through a phenomenon known as the Antarctic Circumpolar Wave. Embedded and slowly moving within the current is a band of water affected by El Niño events in the central Pacific Ocean. This complex southern climate regime continues to befuddle climatologists trained in relatively regular Northern Hemisphere weather patterns.

Southern Hemisphere albatrosses nest on island groups that lie in or close to areas of saltwater convergence (see Map 2). New Zealand owns seven mostly sub-Antarctic archipelagoes in the southwest Pacific: South Island with its offshore Three Kings Islands and Solander Islands, Campbell Island, the Snares, Bounty Island, the Antipodes, the Auckland Islands, and the most easterly ten-island Chathams, bathed by nutrient-rich flows over the continental shelf. Twelve of the twenty-two species of albatrosses, comprising about 250,000 annual breeding pairs, nest on sixteen sites in New Zealand. Eight of these albatrosses nest nowhere else, so New Zealand easily merits its claim as the albatross capital of the world.

On the eastern side of the Pacific Ocean, set along the north shallow-sea perimeter of the Drake Passage between the Pacific and Atlantic Oceans, lie the albatross nesting islands Chile claims off its south coast and fiordland; Isla Diego de Almagro, Islas Ildefonso, and Islas Diego Ramírez are the important ones.

The South Atlantic Ocean west of the Mid-Atlantic Ridge has four groups of albatross nesting islands claimed by Britain under the 2002 British Overseas

Territories Act. The Falkland Islands, about 300 miles east of Tierra del Fuego, include the two main islands, East and West Falkland, and more than seven hundred smaller islands. The Falklands' 4,700 square miles form the largest of the UK's island territories. About 1,000 miles southeast of the Falklands, UK's South Georgia includes the large mountainous main island and several prominent offshore islands, including Bird Island, where Lance Tickell and John Croxall and his British Antarctic Survey colleagues studied albatrosses and continue to do so today. Both archipelagoes are albatross strongholds.

East of the lower half of the Mid-Atlantic Ridge lie the Tristan da Cunha Islands at 37°S latitude, a dependency of the British territory of Saint Helena. Tristan da Cunha consists of the inhabited Main Island and uninhabited ones such as Nightingale, Inaccessible, and Gough, a major breeding site for globally threatened albatrosses.

Moving east, we come across the Prince Edward Islands and the much larger Marion Island, both claimed by South Africa and home to five albatross species. In the south end of the Indian Ocean, France claims a number of islands and archipelagoes as its Terres australes et antarctiques francaises (TAAF), including Iles Crozet, Amsterdam and St. Paul, and Iles Kerguelen. Seven albatross species

Black-browed Albatrosses nesting on southwest-facing slopes on Grand Jason Island have recovered since sheep were removed in the late 1960s, but they face an uncertain future in their stronghold, the Falklands. At-sea losses due to commercial fishing have turned this populous and widespread seabird into an endangered species. Photo courtesy Robin Woods.

breed on these islands. Southeast of Kerguelen lie Australia's most isolated and frigid Heard and McDonald Islands (HIMI), and passing into the Pacific Ocean, at 54.5°S, we reach Macquarie Island, the most southerly albatross nesting island after those claimed by Chile. It is southeast of the State of Tasmania and halfway between that island and Antarctica. Tasmania manages Macquarie Island and its four albatross species, and Australia has jurisdiction over the waters around Macquarie Island from twelve miles out to three hundred miles from shore. Tasmania manages three additional islets off its coast—Albatross Island, Mewstone, and Pedra Branca—that support Australia's single endemic species, the Shy Albatross. Moving east from Macquarie Island brings us full circle back to New Zealand's sub-Antarctic set.

CHAPTER 6

Family

FOSSIL EVIDENCE CONFIRMS that many existing albatrosses and other sea-birds originated in the Southern Hemisphere, where saltwater covers more than two-thirds of the planet's surface. Researchers have discovered albatross-like fossils fifty million years old from the Lower Eocene Epoch that probably share an earlier common ancestry with penguins. Fossils from thirty million years ago in the Oligocene Epoch, and identified more conclusively as albatross-type birds, suggest that these large seabirds flew in all the major oceans. The earliest look-alikes of the four current genera turn up in the fossil record of the Miocene Epoch about twenty or fifteen million years ago. Lance Tickell suggests they may have coevolved with squid, their food of choice. A small number of fossils found in Argentina, South Africa, and Australia dating from nine to five million years ago demonstrate the presence of albatrosses in the Southern Ocean in the Lower Pliocene Epoch. Fossils from five species indicate albatrosses also soared in the North Atlantic Ocean along the coasts of both North America and Europe. It may be that albatrosses evolved separately from each other in the oceans of the Northern and Southern Hemispheres.

After the Panama isthmus separated the Pacific Ocean from the Atlantic Ocean two to three million years ago, albatrosses eventually disappeared from the fossil records of the North Atlantic. When they are sighted there today, people generally think of them as loners or lost birds. The Short-tailed Albatross soared commonly into the Aleutian Islands in the North Pacific, and evidence of predation by indigenous people appears in middens from Alaska into Southern California, where hunters fashioned bones of Short-tailed and other albatrosses into artifacts. There is also evidence of early albatross hunting as Polynesians moved from the tropics into Hawaii and New Zealand.

Albatrosses are obviously seabirds par excellence, but what is a seabird? Fewer than 3 percent of all named bird species, that is, only two or three hundred of the almost ten thousand species, have learned to exploit the marine ecosystems that blanket 70 percent of the globe's surface. These species live in salty oceans and seas and also in coastal marshes, beaches, estuaries, and wetlands. They belong to four orders:

· Sphenisciformes—penguins
· Procellariiformes—albatrosses, shearwaters, and other petrels
· Pelecaniformes—shags, gannets, and their kin
· Charadriiformes—terns, gulls, and skuas.

There is still no definitive agreement, however, about which of these bird species should be called seabirds. We've relied instead on cumulative attempts to define and construct taxonomy, beginning with the work of scientists like Murphy, Matthews, Alexander, and Richdale. When Alexander published his revised edition of *Birds of the Ocean* in 1954, he included Richdale's life histories of New Zealand's rich seabird avifauna, "as prolonged and detailed as any devoted to land birds."[1] Other contributions to the discussion of what a seabird is included James Fisher's *Fulmar* (1952); Nikolaas Tinbergen's *Herring Gull's World* (1953), for which Tinbergen shared a Nobel Prize; and the work of Fisher's collaborator Ronald M. Lockley, who dedicated a long career to popular and scientific writing about Northern Hemisphere auks, petrels, and shearwaters. By the time of Alexander's revised *Birds of the Ocean*, Lockley had published *Shearwaters* (1942) and *Puffins* (1953). Lockley's *Seabirds* (1954), coauthored with Fisher, offered the best summary of what was known at the time. They reviewed research done in European waters and pointed beyond the corridor of the North Atlantic.

Thirty years later, three distinguished bird scientists brought out a 780-page summation of the distribution and abundance of seabirds around the world. They also examined threats to various species and how best to deal with them. The 1984 *Status and Conservation of the World's Seabirds* was the first publication of the Seabird Specialty Group of the International Council for Bird Preservation (ICBP). The ICBP had been working since 1922 through 270 member organizations in eighty-six countries to protect wild birds and their habitats. In the foreword to the book, ICBP director Christoph Imboden spoke of the inspiration we garner from observing seabirds' mastery of the watery domain and emphasized that seabirds serve as signs of the state of the oceans, the planet's global commons.

Edited by British Antarctic Survey seabird authority John P. Croxall, Peter G. H. Evans from the Edward Grey Institute, and Ralph W. Schreiber from the Los Angeles County Museum of Natural History, the book was the product of a three-day workshop and symposium held in Cambridge, England, in 1982, attended by fifty seabird biologists from sixteen nations.[2] It presented the most authoritative argument to date about the structure and members of the seabird family. Admitting a dilemma in regard to what constitutes a seabird, the editors deliberately excluded denizens of inland seas, marshes, and coastal waters, such as seaducks, loons, and shorebirds, some of which, like phalaropes, do pass time at sea. They included only species that feed in coastal waters or the open seas: all albatrosses, petrels, and their allies; auks; skuas; the marine gulls, terns, and skimmers; most cormorants; and some pelicans. They selected 282 species in fourteen families in the four seabird orders.[3]

Marine ornithologist Anthony Gaston further divides seabirds into 222 wholly marine and 72 partially marine species. Albatrosses, petrels, storm-petrels, boobies, frigate birds, tropicbirds, and auks fall into the category of wholly marine seabirds that feed exclusively offshore, usually out of sight of land. Gaston's classification in general aligns with the 1984 *Status and Conservation of the World's Seabirds*. Similarly, together with seaducks, he excludes many gull and tern species because they stick close to shore and have, he argues, little measurable effect on marine resources.

There is no debate about whether albatrosses qualify as seabirds, along with all their Procellariiform cousins, the shearwaters, storm-petrels, and diving petrels. These largest and heaviest flying birds have great ability to retain heat, and their size and stronger skulls compared with land birds equip them to cope with storm winds and waters and to plunge-dive from great heights for food. Within the Procellariiform order, the family Diomedeidae includes all the albatrosses.[4]

The debate is ongoing, however, about how many albatross species there are. Terence Lindsey's 2008 *Albatross* most recently summarizes and discusses this quite complicated debate. For most of the last century, experts agreed on a total of fourteen species. Swedish scientist Carl Linnaeus (1707–1778), who initially brought consistency to naming plants and animals through his binomial system, called the iconic Wandering Albatross, among the largest and most strikingly colored of all seabirds, *Diomedea exulans* in 1758, after Diomedes, king of Thrace, and *exulans*, the Latin word indicating that something is to be exiled or banished. After participating in the siege of Troy, in a later campaign, Diomedes offended the goddess Athena, and she blasted his fleet with a great storm and then turned the king and his men into large white birds. In the 250 years since Linnaeus named the Wanderer, collectors and taxonomists confusingly have used about eighty-five additional names to describe this same bird species, the result both of collecting specimens randomly by shooting at sea and of the range of plumage a single bird may display over the natural breeding cycle as well as over the course of a fifty- or sixty-year life. These factors also led collectors to claim that birds of the same species belonged to completely separate species. Enter DNA-based science, and the picture begins to clear up. Or does it? Through their DNA research, New Zealanders Chris Robertson and Gary Nunn reconfirmed in 1998 a total of four genera within the albatross family.

A controversial aspect of the 1998 DNA analysis, however, suggested that the species within the four generally agreed-upon albatross genera should be almost doubled, from fourteen to twenty-four. This meant raising the status of several populations that nested on specific islands to that of species, not the subspecies level most people had accepted. For example, under the new genetic research, the Wandering Albatross remained a type species for five major island groups where it nests, but populations on additional nest archipelagoes were upgraded to species status and named the Amsterdam Albatross, the Antipodean Albatross, the Tristan Albatross, and Gibson's Albatross. This DNA-based differentiation remains controversial because taxonomically it is based on very little genetic distance or divergence among these new bird species, actually less than 1 percent, whereas for most birds we call species, the amount of genetic difference is at least double that. Furthermore, among these species with minimal genetic differentiation, like the Wandering Albatross and the Amsterdam Albatross, it is not genetics, nor is it body size, shape, color, or bird behavior that keeps the species from interbreeding. It is geography, which may have to become a factor in species definition.

The argument over albatross taxonomy is not merely an argument among scientists about ways of naming discrete elements of the observable world. It has political and economic consequences, too. Conservation programs pour resources into rehabilitating dwindling stocks of plants and animals. So renaming the twenty or so pairs of what were previously thought to be Wandering Albatrosses nesting on Amsterdam Island as Amsterdam Albatrosses led also to their status being defined as Critically Endangered, greatly increasing their visibility and claim on limited conservation resources. By and large, today a taxonomy of four genera and twenty-two species of albatrosses is accepted, most significantly by the International Union for the Conservation of Nature (IUCN) and BirdLife International:

- Great Albatrosses (*Diomedea*): six species, including the Northern and Southern Royal Albatrosses and the Wandering Albatross
- North Pacific Albatrosses (*Phoebastria*): four species, including the Laysan, Short-tailed, and Black-footed Albatrosses. During the breeding season, the Waved Albatross inhabits waters off the Galápagos Islands, where most birds nest on a single island.
- Mollymawks (*Thalassarche*): ten species, including peripatetic birds that fly in Northern Hemisphere waters such as the Black-browed and Yellow-nosed Albatrosses.
- Sooty Albatrosses (*Phoebetria*): two species, the Sooty and the Light-mantled Albatross.

FIGURE 1. Albatross Taxonomy. After Robertson and Nunn 1998.

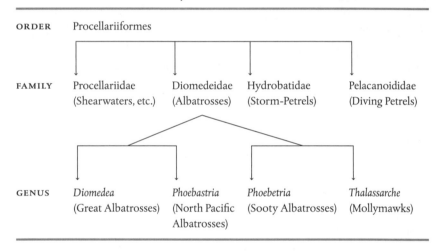

ORDER	Procellariiformes			
FAMILY	Procellariidae (Shearwaters, etc.)	Diomedeidae (Albatrosses)	Hydrobatidae (Storm-Petrels)	Pelacanoididae (Diving Petrels)
GENUS	*Diomedea* (Great Albatrosses)	*Phoebastria* (North Pacific Albatrosses)	*Phoebetria* (Sooty Albatrosses)	*Thalassarche* (Mollymawks)

It is not only the albatross's magnificent appearance that captures our imagination and evokes our awe. We also find their life history fascinating, including their marvelous and faithful feats of navigation and homing. After the first jump into the air that a fledgling successfully combines with forward momentum, it glides out to sea, where it begins to flap then glide, instinctively learning to look for and find food, spending three, six, or even more years away from land before returning to its birth colony. Guided by hormones, memory, experience, and smell, the maturing bird returns to the site where it hatched—even to the same grass and nest pedestal that its parents constructed years before—to begin several years of seasonally associating with other adolescent birds in dazzling aerial displays and courting in elegant dancelike rituals to find a mate. We don't know exactly how the albatross solves the riddle of homing across enormous stretches of ocean to its island birth site after years of circling oceans and seas far from land.

Once an egg is laid, the partners take turns sitting on the nest and flying out to sea to forage. One will leave at first for a few days and then return to relieve its mate from guard duties. About a month after the chick hatches, the parents begin leaving together, making the chick fend for itself for longer and longer spans of time. Nest building to fledging takes several months and up to a year or more for the species of Great Albatrosses. Foraging places great pressure on adults. They have to fuel themselves for the round trip while traveling thousands of miles to gather food for their chick. Then, after successfully raising a chick that fledges, in due course, perhaps after one year, two, or even longer out at sea again, depending on the species and individual bird preferences, the adults return to each other at about the same time to the same site to begin another nest season together.

GREAT ALBATROSSES (*DIOMEDEA*)

The evocatively named Wandering Albatross is the quintessential albatross—large, eye-catchingly colored with cream bodies and dark wings, peripatetic, and, to us, slightly menacing in appearance. They inhabit a band of latitude some forty-five degrees wide in the Southern Ocean, foraging in summer along the edge of the Antarctic pack ice and heading northward in winter into waters along the Tropic of Capricorn as far as the Patagonian Shelf off Chile and to the coast of central Namibia off West Africa. Some fly into waters within ten degrees of the equator.[5] Successful Wanderer pairs take flying sabbaticals of at least one breeding season after fledging a chick before returning faithfully to their nest island, often using the same nest site.

Wanderers inhabit the waters of twelve nations and ten regional fisheries bodies. They nest in many places on South Georgia, where Murphy, Matthews, Tickell, and Croxall encountered them. Eighty percent of all Wanderers nest on four island groups in the Indian Ocean sector of the Southern Ocean, claimed by South Africa and France, where populations intermingle along the eastern edge of the sunken Kerguelen shelf around Kerguelen and Crozet and over seamounts between the Marion and Prince Edward Islands. Reliable information exists for about eight populations, and estimates around the turn of this current century placed the world total at about eight thousand breeding pairs or 28,000 mature birds (Table 1).

ABOVE: First described by Linnaeus (1758), the Wandering Albatross is among the large Great Albatrosses, and has recently been separated into five similar species based on plumage and geographical characteristics. BELOW: Wandering Albatross pair in courtship display consisting of sky-pointing, calling, and wing extensions. Numbers on South Georgia continue to fall. Courtesy Greg Lasley.

TABLE 1. Status of the Albatrosses

Genus *Diomedea*: Great Albatrosses are the largest and longest-lived albatrosses; they are currently separated into 6 species with a complex phylogeographic pattern experts continue to debate.

SPECIES NAME	POPULATION	NEST LOCATION: ANNUAL PAIRS	STATUS
Southern Royal Albatross (*Diomedea epomophora*)	15,000 pairs, about 50,000 birds	**NZ:** endemic; Campbell: 7,800; Auckland (Enderby): 70	Vulnerable
Northern Royal Albatross (*D. sanfordi*)	ca. 5,800 annual pairs, 17,000 mature birds	**NZ:** endemic; Chatham (3 islands); South Island; Taiaroa Head: 30	Endangered
Wandering Albatross (*D. exulans*)	8,100 annual pairs, approx. 28,000 mature birds	**France:** Kerguelen Island: 1,100; Crozet: 2,000 **South Africa:** Prince Edward Islands: 3,500 **UK:** South Georgia: 1,400 **Australia:** Macquarie Island: 8	Vulnerable
Antipodean Albatross* (*D. antipodensis*)	23,500 mature birds	**NZ:** endemic; Antipodes Islands: 4,600 pairs; Auckland Islands: 5,800 pairs	Vulnerable
Tristan Albatross (*D. dabbenena*)	9,000–15,000 mature birds	**UK:** endemic; Only Gough Island: 1,640 pairs	Critically Endangered
Amsterdam Albatross (*D. amsterdamensis*)	25 annual pairs, approx. 100 mature birds	**France:** endemic; Amsterdam Island: 25	Critically Endangered

Genus *Phoebastria*: North Pacific Albatrosses are the closest relatives of *Diomedea* (splitting at least 15 million years ago). They are medium-sized and geographically isolated from Southern Hemisphere cousins. Three species inhabit the North Pacific, plus the Waved Albatross, which nests mostly on the Galápagos Islands.

SPECIES NAME	POPULATION	NEST LOCATION: ANNUAL PAIRS	STATUS
Laysan Albatross (*Phoebastria immutabilis*)	590,000 pairs, approx. 1.2 million mature birds	**US territories** (breeds on 13 sites): Midway: 396,000; Laysan: 142,000; Lisianski: 26,000; Kure Atoll: 14,000 **Japan** (Bonin Islands): 20 **Mexico** (4 sites): 500	Near Threatened

SPECIES NAME	POPULATION	NEST LOCATION: ANNUAL PAIRS	STATUS
Black-footed Albatross (*P. nigripes*)	ca. 129,000 mature birds	**US territories** (breeds on 10 sites): Midway: 24,000; Laysan: 19,000 **Japan:** Torishima: 2,150; Plus two archipelagoes	Endangered
Waved Albatross (*P. irrorata*)	15,000 pairs, 70,000–80,000 mature birds	**Ecuador:** endemic on Galápagos: 10,000	Critically Endangered
Short-tailed Albatross (*P. albatrus*)	1,026 pairs, 2,360 mature birds	**Japan:** endemic; Torishima: 418; Senkaku: 52	Vulnerable

Mollymawks are a medium-sized, diverse group of about 10 species circumpolar in the Southern Hemisphere. They are identified by light and dark heads and bill color patterns, including bare flesh.

SPECIES NAME	POPULATION	NEST LOCATION: ANNUAL PAIRS	STATUS
Atlantic Yellow-nosed Albatross (*Thalassarche chlororhynchos*)	ca. 33,250 pairs, 55,000–83,000 mature birds	**UK territories:** endemic; Gough: 5,000; Tristan da Cunha: 16,000–30,000	Endangered
Indian Yellow-nosed Albatross (*T. carteri*)	ca. 41,500 pairs, ca. 160,000 mature birds	**French territories:** Amsterdam: 27,000; Crozet: 7,000 **South Africa:** Prince Edward Island: 7,000	Endangered
Buller's Albatross (*T. bulleri*)	30,500 pairs, approx. 50,000–55,000 mature birds	**NZ:** endemic; Snares: 8,700; Solander: 4,900; Chatham—Forty-Fours: 16,000	Near Threatened
Shy Albatross (*T. cauta*)	12,750 pairs, 50,000–60,000 mature birds	**Australia:** endemic; Tasmania: Albatross: 5,100; Mewstone: 7,300; Pedra Branca: 270	Near Threatened
Chatham Albatross (*T. eremita*)	4,575 pairs, 18,000–20,000 mature birds	**NZ:** endemic; Chathams (Pyramid Rock): all nests	Critically Endangered

SPECIES NAME	POPULATION	NEST LOCATION: ANNUAL PAIRS	STATUS
Salvin's Albatross (*T. salvini*)	32,000 pairs	**NZ:** endemic; Bounty: 30,700; Snares: 1,200	Vulnerable
Campbell Albatross (*T. impavida*)	ca. 50,000 mature birds	**NZ:** endemic; Campbell: 21,000	Vulnerable
Grey-headed Albatross (*T. chrysostoma*)	92,300 pairs, approx. 250,000 mature birds	**UK territories:** South Georgia: 47,000 **French territories:** Kerguelen: 7,900; Crozet: 6,000 **Chile:** Diego Ramírez: 17,000; **NZ:** Campbell: 6,600 **South Africa:** Prince Edward Islands: 10,300 **Australia:** Macquarie: 84	
Black-browed Albatross (*T. melanophrys*)	530,000 pairs, ca. 2 million mature birds, being revised downward	**UK territories:** Falklands (12 islands): 399,000; South Georgia (20 sites): 74,300 **Chile** (6 islands): 123,000 **France:** Crozet: 980; Kerguelen: 3,200 **Australia:** 780 **NZ:** 135	Endangered
White-capped Albatross** (*T. steadi*)	97,000 pairs, ca. 360,000 mature birds	**NZ:** endemic; Auckland: 97,000	Near Threatened

Sooty Albatrosses are the closest cousins to *Thalassarche* birds (splitting from Molly-mawks at least 10 million years ago). They are agile and vocal, medium to small, brown in two species: one is circumpolar and disperses across the Southern Ocean; the other forages over lower latitudes in the same hemisphere.

SPECIES NAME	POPULATION	NEST LOCATION: ANNUAL PAIRS	STATUS
Sooty Albatross (*Phoebetria fusca*)	ca. 13,300 pairs, approx. 42,000 mature birds	**UK territories:** Gough: 5,000; Tristan da Cunha (3 islands): 7,500 **France:** Amsterdam: 470; Crozet: 2,100 **South Africa:** Prince Edward Islands: 2,500	Endangered

SPECIES NAME	POPULATION	NEST LOCATION: ANNUAL PAIRS	STATUS
Light-mantled Albatross (*P. palpebrata*)	22,031 pairs, approx. 140,000 birds	**UK territories:** South Georgia: 5,000 **South Africa:** Prince Edward Islands: 650 **France:** Kerguelen: 4,000; Crozet: 2,300 **Australia:** HIMI: ca. 350; Macquarie: 1,280 **NZ:** Auckland: 5,000; Campbell: 1,600; Antipodes: 250	Near Threatened

* Species under debate; First Meeting of Parties to ACAP (2004) stipulated that a Working Group on Taxonomy should report to the Advisory Committee regarding the split between *D. gibsoni* and *D. antipodensis*. In 2008, ACAP listed Antipodean (*D. antipodensis*) as a separate species.

** Species under debate by ACAP Working Group regarding the split between *T. cauta*/ *steadi*, and *T. bulleri/platei*. In 2008, ACAP listed Shy (*T. cauta*), White-capped (*T. steadi*), and Buller's (*T. bulleri*) as separate species. Discussion has centered on the Shy group that split into Salvin's and Chatham in 2004 and the White-capped in 2008.

Sources: ACAP 2010b; BirdLife International 2010a; IUCN 2010.

Long-term research on South Georgia's Bird Island has supplied information about migration and movements of the Wandering Albatross as well as accurate but disturbing data about downward population trends. BirdLife International estimates that the number of Wandering Albatrosses has declined 30 percent or more over a three-generation span of seventy years. As the official listing authority for the IUCN's Red List of threatened birds, BirdLife designates the Wanderer's status as Vulnerable and has been unable to document substantial recovery.

The other Great Albatrosses, the Northern and Southern Royals, which some have questioned as separate species, since they do hybridize, are larger even than many Wanderers. Circumpolar as ocean foragers, 99 percent of all Northern Royal Albatrosses nest on three small privately owned islands in New Zealand's Chatham Island chain. This is the albatross Richdale devoted his life to studying, and it is probably the most well known among birding enthusiasts because of the accessible colony on Taiaroa Head within the city limits of Dunedin. Although the number of nesting pairs on Taiaroa Head (less than 1 percent of breeding birds) has increased, overall decreases of both Northern and Southern Royal Albatrosses began to be noted in the 1980s, despite earlier evidence of their recovery from sealing and whaling and the habitat destruction of farming and introduced animals and plants. A severe storm on the Chatham Islands in 1985 caused serious vegetation loss and erosion in the nesting colonies, forcing birds to lay their eggs on virtually bare rock with minimal nest lining. Dramatic failure in nesting, notably in the Forty-Fours where two-thirds of breeders nest, as well as the fact that in the off-season these birds fly east more than six thousand miles into the Patagonian Shelf off Argentina, which is a key commercial fishing area, has turned this New Zealand endemic albatross into an endangered species. In the years following the storm, only about one pair in five successfully fledged a chick in the Chatham Island colonies. Recently, fledging success has risen to about two-thirds of the prestorm rate as herb and grass cover on very thin soil has reappeared in nest colonies. Scientists estimate that there are maybe 34,000 Northern Royal Albatrosses and 5,800 breeding pairs (see Table 1).

Virtually all breeding Southern Royal Albatrosses nest on New Zealand's Campbell Island. Considered to be Vulnerable rather than Endangered, numbers appear to have increased over the past twenty-five years, although experts note a leveling off. The nonbreeding Southern Royals also favor the Patagonian Shelf for foraging, which greatly increases their risk of being hooked in commercial fishing. Current estimates place the Southern Royal population size at about 50,000 birds, including 6,500–7,000 annually breeding pairs (see Table 1).

PACIFIC ALBATROSSES (PHOEBASTRIA)

Three Pacific albatrosses—the Laysan, Short-tailed, and Black-footed—nest exclusively in the Northern Hemisphere on twenty-four sites, about half of which are located on a string of islands in the central Pacific northwest of Hawaii (see Map 1). The fourth, the Waved Albatross, nests in the Galápagos Islands. Almost 70 percent of the estimated 590,000 breeding Laysan pairs nest in colonies on Midway Atoll, another 24 percent nest on Laysan Island, and seven additional sites have more than one hundred pairs. Recently, some Laysans have started to breed on islands claimed by Japan and Mexico and also on the three main islands of Hawaii: Kauai, Oahu, and Niihau. The creation in 2006 of US Papahānaumokuākea Marine National Monument, now a World Heritage Site, which spans all the northwestern Hawaiian Islands, provides extra protection for seabird colonies.[6]

Despite the large population of Laysan Albatrosses, which increased after the terrible slaughter by the feather merchants a century ago, the numbers of Laysans have tapered off. Recent estimates suggest that the nesting population

Short-tails were declared extinct in Japanese nest areas but have made a spectacular return on their home base of Torishima Island, south of Tokyo. Courtesy US Fish and Wildlife Service.

within the US National Wildlife Refuge system, which includes Midway Atoll, and in sanctuaries administered by Hawaii declined by one-third during the 1990s. Mortality is also high from the birds' ingestion of floating plastic, trash, and lead paint chipped by chicks off buildings on Midway and oil pollution. The stomach of one six-month-old Laysan chick that starved to death contained well over 250 items its parents fed it from their stomach contents, including two cigarette lighters, a shotgun shell, a pump sprayer, clothespins, bottle tops, and many large and small bits of plastic such as plastic toys and toy parts. The high Laysan mortality rate has led officials to classify the Laysan Albatross as Vulnerable.[7]

The IUCN has changed the status of the Black-footed Albatross (*P. nigripes*) from Vulnerable to Endangered since 2000 (see Table 5). There are currently about 64,000 pairs and almost 280,000 birds in the North Pacific, nesting primarily on Midway, Laysan, and French Frigate Shoals (see Map 1). And as we noted in reporting Hiroshi Hasegawa's devoted work to restore the population of Short-tailed Albatrosses on Torishima and now on the Bonin Islands, Short-tails are increasing annually on Torishima and also on Senkaku (see Map 1). As a result, the IUCN status for the Short-tailed Albatross is currently Vulnerable. It is estimated that there are about 1,000 breeding adults, 85 percent of which nest on Torishima. Hasegawa's efforts are paying off.

Although the international environmental regime that formed eventually to save the albatross, the subject of this book, elected to focus on the Southern Hemisphere first, it now has spread its jurisdiction worldwide. This will allow integration into the global picture of the years of knowledge gained by intrepid scientists like Harvey and Mildred Fisher, working to study and save the Laysan Albatross on Midway Atoll, and Hiroshi Hasegawa, working to save the Short-tailed Albatross on Torishima.

MOLLYMAWKS (*THALASSARCHE*)

Fishers and seamen in New Zealand, the Falkland Islands, and Tristan da Cunha named the mid-sized, mostly black and white albatrosses they saw Mollymawks, a word derived from the Dutch for "foolish birds."[8] Mollymawks sometimes get into difficulty as they swoop fearlessly into offal and the flesh of dead seals and whales. These well-distributed Southern Hemisphere seabirds comprise ten species, the largest of which are the Black-browed and the Yellow-nosed. The Black-browed Albatross is the most globally widespread representative of all albatrosses. Until recently, it was also the world's most abundant albatross, but the Laysan replaced it at the beginning of the twenty-first century. Mollymawks are most closely related to the Sooty Albatrosses, and their total population tops 2.5 million, including about one million breeding pairs.

Black-browed Albatrosses range widely throughout the Southern Hemisphere and nest on ten island groups belonging to five nations. They forage in sub-Antarctic waters close to nest islands during the southern spring and summer, from August to April, most abundantly around four South Atlantic Ocean nest sites, especially the Falkland Islands, where a dozen colonies hold about two-thirds of the world's breeders. After maturing for three to eight years at sea, young adults head back to their natal islands to court and begin to breed at six to thirteen years of age. After nesting, adult Black-brows head northward, also exposing themselves to fishing activities.

Satellite tracking suggests that Black-browed breeding populations remain more or less distinct. Birds from South Georgia, which tend to work northwest over the Patagonian Shelf, do not mingle with birds that breed in Chile, which may fly as far as New Zealand waters to capture squid along the convergence zones. Although individuals intermingle with others of the same species in winter when they forage over longer distances, they still return to their separate nesting colonies for the next breeding cycle.[9]

ABOVE: Black-browed Albatrosses are among the most numerous but are now endangered due to huge losses sustained by the Falkland Islands birds that forage mainly in the southern Atlantic Ocean and get taken by fishers. Courtesy Greg Lasley. BELOW: The Grey-headed Albatross typifies extensive ocean travels, exemplified by the round-the-world journeys taken by two males tagged on South Georgia, a nest island. Courtesy Greg Lasley.

In 2003, because of the Black-brow's calculated sixty-five percent population decline over a sixty-five-year period, BirdLife International listed it as Endangered worldwide. Massive losses have been noted in the last twenty years. From 2000 to 2005 alone, some researchers estimate that as many as a staggering 80,000 breeding pairs disappeared from the Falklands, its nesting stronghold. Recent estimates place the world population of breeding Black-browed Albatrosses at about 530,000 breeding pairs and a total population that BirdLife estimates to be just over one million mature birds (see Table 5).

Mollymawks occasionally show up in quite unexpected places. In 1860, a Black-browed Albatross arrived in a gannetry near Myggenaes Holm on the Faeroes, to which it returned faithfully every February for thirty-four years before someone shot it. Locally named King of the Gannets, it turned out to be a female and found a final resting place in the Copenhagen Museum. An exhausted Black-browed Albatross also turned up near Linten, Cambridgeshire, England, in 1897. Over the following century, observers recorded twenty-seven Black-browed Albatrosses in waters around Britain and Ireland, including two that had arrived inland. In the 1960s, birders also spotted Black-browed Albatrosses off Cornwall, Devon, and Yorkshire. Eleven of the sightings occurred between 1963 and 1968, leading some to speculate that Black-brows were moving farther north in the Atlantic Ocean, but this tally may be the result of a growing interest in seabird watching. Like the earlier King of the Gannets, a Black-browed Albatross hung out in a similar gannetry on Bass Rock, East Lothian, from May through September 1967, and again from April through July 1968. Another adult, nicknamed Albert Ross, summered on Hermaness in the Shetlands in 1972 and returned annually through 1987, then again from 1990 through 1994. Enthusiasts expressed high hopes that a new seabird was about to nest in Britain. However, in all those years of faithful return, Albert Ross never attracted a mate and he finally disappeared.[10]

In 2007, another Mollymawk, a Yellow-nosed Albatross, landed on a resident's driveway in Brean Sands, Somerset, England, barely twenty miles from Watchet, the hearth of Coleridge's ancient mariner. Cared for by a wildlife rescue center near Burnham-on-Sea, after a night's rest, the bird lifted off and made out to sea again. But it must have done an about-face, as a day later it was seen close to the North Sea in a fishing lake in Lincolnshire and a few days later turned up near Malmö, Sweden. The Rarities Committee accepted this as a genuine vagrant from the South Atlantic and therefore new to Britain. Yellow-noses tend to be better known on the warmer side of North America, where people reported four sightings in the early 1970s along the Texas Gulf Coast's South Padre Island.[11]

Albatrosses other than the globally adventurous Black-browed and Yellow-nosed have made their way into the North Pacific Ocean from the Southern Hemisphere. Shy Albatross sightings have been reported in Washington State and off the coast of Oregon and California's Mendocino Coast, the most recent in 2000. In addition, in recent years observers have also reported three Salvin's Albatrosses off California and another near the Aleutians. Whether these birds are beginning to represent a tiny semiresident population, as some have begun to suggest, is open to speculation. The number of reported sightings is increasing.

SOOTY ALBATROSSES (PHOEBETRIA)

Sooty Albatrosses nest on fifteen sites on six archipelagoes in the Southern Hemisphere: Tristan, Gough, Prince Edward, Crozet, Kerguelen, and Amsterdam. Adolescents may remain at sea for eight years before homing to their breeding islands, and they don't pair up and begin to nest until they are about twelve years old. Sooties disperse widely to forage in the South Atlantic and South Indian Oceans. There are about thirteen thousand annual breeding pairs, but their numbers, like those of so many other albatross species, are declining, and their IUCN status is Endangered.

Light-mantled Albatrosses are the most elegant species; they use ledges as nest sites, forage into polar waters, and can dive thirty feet or more below the surface to capture food. This one is from South Georgia. Courtesy Greg Lasley.

Similar trends exist for the second member of the Sooty genus, the Light-mantled Albatross, which nests on nine islands in the Southern Hemisphere and ranges widely at sea, moving into the Antarctic pack ice farther south than any other albatross and as far north as the coastal waters of Peru. One of these stunningly beautiful small light gray–brown albatrosses was sighted in 1994 near California's offshore birding hot spot, Cordell Banks. The Light-mantled population is estimated to comprise close to double the number of Sooty breeding pairs, or twenty-five thousand pairs, and it enjoys the more favorable Near Threatened IUCN status.

Identifying these downward trends in the overall population of the albatross family after the 1960s and their causes was another piece of global scientific detective work, carried out at first independently and in widely separated spaces by French, Australian, New Zealand, South African, and British scientists, and then collaboratively as they began to put pieces of the circumpolar puzzle together. The discovery of a declining trend in the 1970s was especially disheartening to experts and enthusiasts worldwide, for at the very moment they began to identify the trend, they all thought albatrosses for the most part had successfully recovered from the damage done to them in the era of unregulated hunting in the Southern Ocean.

First described by Linnaeus (1758), the Wandering Albatross is among the large Great Albatrosses and has recently been separated into five similar species based on plumage and geographical characteristics. Courtesy Greg Lasley.

ABOVE: Wandering Albatross pair in courtship display consisting of sky-pointing, calling, and wing extensions. Numbers on South Georgia continue to fall. Courtesy Greg Lasley.
BELOW: Short-tails were declared extinct in Japanese nest areas but have made a spectacular return on their home base of Torishima Island, south of Tokyo. Courtesy US Fish and Wildlife Service.

ABOVE: Black-browed Albatrosses are among the most numerous but are now endangered due to huge losses sustained by the Falkland Islands birds that forage mainly in the southern Atlantic Ocean and get taken by fishers. Courtesy Greg Lasley. BELOW: Light-mantled Albatrosses are the most elegant species; they use ledges as nest sites, forage into polar waters, and can dive thirty feet or more below the surface to capture food. This one is from South Georgia. Courtesy Greg Lasley.

ABOVE: Laysan Albatrosses have flourished on Midway Atoll since the US military left; however, longlines, plastic objects fed to chicks from the "garbage patch" northwest of the atoll, and ingested lead paint take their toll. Here Laysan Albatross nests are being counted. Courtesy US Fish and Wildlife Service. BELOW: An example of a dead albatross chick compacted with floating objects taken as food. Photo © Chris Jordan.

ABOVE: Button fathered the five-hundredth chick, named Toroa, recorded on Taiaroa Head, Dunedin, and was the last male born to "elderly" Grandma in 1989. Hand-fed in the nest, he is a regular member of the New Zealand colony. Courtesy Isobel Burns and Lyndon Perriman. **BELOW:** Trawlers attract birds, which strike the cables supporting the net, or are trapped in fish-filled meshes as crewmembers haul it aboard. This is an ongoing research problem. Photo by Robert Grace, © Greenpeace. Courtesy of Greenpeace.

ABOVE: Grytviken Whaling Station was a port for Murphy, Matthews, and others working on South Georgia. The hulks of two whale catchers reflect the era of marine mammal and bird exploitation. Courtesy Greg Lasley. BELOW: Short-tailed Albatrosses are being transported from Torishima to Mukojima on the Bonin Islands, a previous nest site. Dummy birds and recorded calls assist these hand-fed chicks to familiarize themselves with the new location and, after several years, return and begin to nest on the island. Courtesy Lyndon Perriman.

The Grey-headed Albatross typifies extensive ocean travels, exemplified by the round-the-world journeys taken by two males tagged on South Georgia, a nest island. Courtesy Greg Lasley.

Wandering Albatross, a casualty of industrial fishing with longlines. Courtesy Graham Robertson.

Part Two

CROSSINGS

CHAPTER 7

Migration

TICKELL AND GIBSON'S DISCOVERY that the albatross is a highly skilled global navigator challenged the assumption that national and local efforts would sufficiently protect and manage albatrosses. Migratory birds—or, for that matter, people, other mammals, fish, insects, plants, minerals, water, and air—that cross political and economic boundaries dramatize the limits of national jurisdictions and authorities. However, to date, no international law of any kind can ultimately be enforced except by the nation-state. Multilateral and international agreements may be binding, but enforcement depends on the will and resources of participating national parties. As a result, the conservation of albatrosses is a matter for intergovernmental agreements and cooperation, and its effectiveness depends on the political will and energy of the twenty-five or more nations on whose territory and in whose waters albatrosses nest and forage, especially the seven nations where most of the Southern Hemisphere species breed.

In the 1850s, as seal and whale gangs in the Southern Ocean decimated mammal and seabird colonies on their breeding islands, Northern Hemisphere nations at first didn't think that limits existed for what they treated as boundless resources for the taking. However, as hauls of mammal skins and oil dropped off, some entrepreneurs and governments began to reckon that actions were necessary to protect and manage animals instead of just exploiting them, something the Maori had apparently known when they selectively harvested albatrosses centuries before. An awareness of limits in regard to wildlife among industrialized nations began well before they recognized similar limits in regard to nonliving resources like gold, oil, water, and air.

A hesitant effort to protect wild birds took place in 1868 at the twenty-sixth General Assembly of German farmers and foresters in Vienna. Participants urged the Austro-Hungarian government to protect species that fed on crop-damaging insects and to seek agreements with other governments about protecting such "useful" bird species. The foreign minister took up the resolution. In so doing, he acknowledged the incipient concept of an ecosystem by linking the well-being of birds to the health of crops, a concept that now, 150 years later, provides a framework crucial to the effectiveness of efforts to manage bird and fish populations. As we shall see, the health of the albatross family is now linked with the health of some of the world's highest-value fish.

The ecosystem approach to resource management is one of three principles that over time have become integral components of environmental discussions, negotiations, international standards, and hard and soft laws. Significant conservation efforts today must address in one way or another the principles of ecosystem management, intergenerational equity, and a precautionary approach to resource management and use. Each one plays its part in the international environmental protection regimes that have developed for the albatross and the fish.

Twenty-one years after the Vienna meeting, in 1889, a group of women near Manchester, England, organized a Society for the Protection of Birds. With backing from the Royal Society for the Prevention of Cruelty to Animals, which enjoyed Queen Victoria's patronage, and similar groups opposing the cruel treatment of animals, this protection movement aimed to enlist women "to unite in discouraging the enormous destruction of bird life exacted by milliners and others."[1] Alfred Newton, first professor of zoology and comparative anatomy at Cambridge University and a founding member of the British Ornithologists' Union, figures prominently in this antiplumage movement, having opposed as early as 1868 killing seabirds for their wings and feathers as they sat on their nest ledges on the chalk cliffs of England's East Yorkshire. While acknowledging utilitarian arguments, public perspectives in regard to bird protection began more broadly, encouraged by Newton, to include humane, aesthetic, and scientific values, especially for species brought to the edge of extinction by the millinery trade. The Royal Society for the Protection of Birds received support from prominent and public persons, including Queen Alexandra, who gave it royal patronage in 1906. In 1910, it joined forces with a similar fledgling group, the Audubon Society, established in the United States in 1886 to press for laws to end sport and commercial killing of wild birds, and the trade, consumption, and wearing of wild-bird plumage. The efforts of Alfred Newton and the Royal Society for the Protection of Birds are important for the albatross because they began to undermine material use of birds in favor of protecting them for their beauty and for their services to human beings. Birds on hats were of no use to farmers. By killing off species that fed on crop-damaging insects, millinery interests were upsetting a "balance of nature," with economic as well as ecological repercussions.

Legislators in Europe and North America grew interested in expanding the incipient body of law that protected wild birds to include setting aside reserves and parks for useful species, including migratory ones. In 1905, a Convention for the Protection of Birds Useful to Agriculture went into force, signed by a dozen nations and covering 150 species deemed helpful to farmers. The convention called upon nations to protect beneficial birds from being hunted during the breeding season. This initial protection act suffered from vague wording and escape clauses and was weakened when Russia, the United Kingdom, and Italy, nations with strongly migratory bird populations, refused to sign. Italy and the island of Malta today are still de facto opposed to prohibitions on hunting during the breeding and migration seasons. They continue to be a target for concerted efforts to limit hunting and to stop liming, a means of trapping small migrant birds for gourmet restaurants and collectors by coating twig perches with glue, which continues on Cyprus and Malta today.[2]

The 1905 convention did show, however, that European nations could initiate and develop international environmental agreements. But these agreements were limited to species useful to people, without regard for other roles birds have in ecosystems. They also lacked any consideration of the affective or ethical range of values articulated by Newton twenty years earlier in regard to birds. In 1916, the United States and Canada signed the first non-European bilateral convention for the protection of birds, implemented as the Migratory Bird Treaty of 1918. It, too, followed utilitarian arguments and established hunting seasons like its European counterpart. By asserting that all migrating species were fully protected, the treaty intended to put a stop to commercial hunting for meat and plumage. Through an executive order more than a decade earlier, Theodore Roosevelt had declared a reserve for a Brown Pelican nest island in Florida, setting in motion what blossomed into a series of about 550 National Wildlife Refuges in the United States today, covering close to 100 million acres. The US Department of Interior's Fish and Wildlife Service manages them as habitats for animals, including migratory waterfowl. In the same year, after complaints about poaching of its seabirds, Roosevelt placed Midway Atoll, annexed by the United States in 1867, under the US Navy, and a year later authorized a detachment of marines to man the atoll in order to keep out Japanese hunters who were slaughtering albatrosses for their feathers. The Fish and Wildlife Service continues to protect wildlife on this atoll and to sponsor ongoing research in permanent stations.

Starting in the mid-1930s, the United States negotiated what became a series of additional bilateral conventions for the protection of migratory bird species flying beyond its borders. The first one, with Mexico, emphasized birds as sources of food and sport, but later agreements, such as a 1972 agreement with Japan, included the scientific and aesthetic in addition to economic values of birds. This trend, to push beyond material considerations, has continued as an increasingly interested and informed public gives voice to an array of values that are more and more seen as interrelated, from inspiration and aesthetics to economics, science, and healthy ecosystems. People draw on these values in their arguments for saving the albatross, including the seemingly conflicting values of fishing efficiency and innovation and fish and seabird sustainability.

In 1940, the member states of the Pan American Union, made up of most of the Central and South American nations, signed an agreement to preserve wildlife in the Western Hemisphere that advanced the cause of bird protection in remarkable ways. It identified human activities causing population declines, established protected areas, regulated trade in wildlife, and paid special attention to migratory species, stressing the need for international cooperation to ensure their protection and management. The preamble was visionary. Parties were

to protect natural habitats of all indigenous species "in sufficient numbers and over areas extensive enough to assure them from becoming extinct through any agency within man's control."[3] Such goals proved ambitious. The agreement did not clearly cover the coastal and saltwater haunts of seabirds, but its intent was to set aside habitats as parks, reserves, and monuments in which a rich diversity of New World bird species could be protected, and to promote the convention's objectives through scientific collaboration. Unfortunately, because the agreement failed to establish an administrative structure to enforce its provisions and monitor progress, it languished as a "sleeping treaty." But even sleeping treaties play an important role in identifying threatened species and their threats, articulating mitigation and conservation standards, and urging people to reassess cultural and economic attitudes. The thinking behind this 1940 convention expanded regard for migratory animals and their status in response to human activity and served as a harbinger for subsequent treaties that have a direct and significant bearing on the albatross family.

In October 1950, an International Convention for the Protection of Birds formalized a growing worldwide recognition that all wild birds, including migratory birds, needed protection from human predation and destruction "in the interests of science, the protection of nature and the economy of each nation."[4] It encapsulated ideas contained in previous bilateral agreements that dealt with the transboundary character of migratory species. It argued that the people of one nation should not destroy a species or group protected by another. Moreover, hunting seasons in all participating nations should be adjusted to accommodate the breeding seasons of different game birds in order to make recreational as well as subsistence hunting sustainable. The 1950 convention was enacted in 1963 after seven European nations had agreed to its provisions. Fatally, however, the agreement included no list of species to protect. Moreover, the absence of reporting and regular meetings of members turned this convention into another sleeping treaty. In any case, carrying out the ambitious objectives of this convention required something that at the time was sorely lacking: accurate scientific knowledge of habitats and population trends, as well as of migratory schedules and pathways of the species that nested or passed through the signatory countries. These weaknesses notwithstanding, this convention, like the Pan American Union agreement a decade before, played an important role in recognizing and expanding nations' responsibilities to all birds in their national territories, whether indigenous or just passing through.

It was not until the 1971 Ramsar Convention on Wetlands in Iran and the 1979 Bonn Convention on the Conservation of Migratory Species of Wild Animals (CMS) that the recognitions and terms in these sleeping treaties took on new life and strength in the form of international soft laws and binding agreements relating to the protection of migratory species. The Ramsar Convention approached the world's wetlands as the complex ecosystems they are, acknowledging the essential services they provide, such as water storage, buffering against flood and coastal storm damage, photosynthesis, filtering and de-nitrification of natural and agricultural runoff, and as habitats for a diversity of plant and animal life. Because of these services, wetlands especially dramatize the need for an ecosystem approach in environmental management. In codifying this approach, the Ramsar agreement added yet another essential layer of foundational material for the development of an effective environmental regime for albatrosses and fish.

During the 1970s, the United States, USSR, Canada, Mexico, Japan, Australia, and India drew up several conventions, mostly bilateral, to protect birds on the basis of a cluster of aesthetic, scientific, and ecological values. Agreement language began to stress the importance of biodiversity and make specific declarations regarding the need to protect endangered species. By the 1970s, using the slowly accumulating body of scientific knowledge, nations, nongovernmental organizations, and individual scientists were also making significant progress in restoring bird habitats, including those of albatross nesting islands in the Southern Ocean. As mentioned above, observations indicated that something seemed to be reversing the numerical recovery of albatross populations after the sealing and whaling devastation. Participants in other environmental conventions of the 1970s began to recognize the dearth of reliable information and data based on systematic investigations over time in order to make effective environmental decisions, an urgent need in regard to species being identified as threatened.

The 1979 Bonn Convention, as the Convention on the Conservation of Migratory Species of Wild Animals is sometimes called, is an intergovernmental agreement aimed at managing and protecting wildlife and wildlife habitats across the globe. As such, it has served as the most far-reaching and effective foundational convention for subsequent agreements for the protection and conservation of albatrosses. As an example of the complexity of environmental agreements, of how interrelated, incremental, and cascading they inevitably are, consider the development of this particular agreement. It was initiated in response to Recommendation 32 of the Action Plan developed at the 1972 UN Conference on the Human Environment in Stockholm. Recommendation 32 called for governments to pay attention to the need for protecting species in international waters and species that migrate across national boundaries for nesting and wintering. After the 11th General Assembly of the World Conservation Union endorsed the proposal, the Federal Republic of Germany took up the initiative to draft a convention for migratory animals. Western Germany was anxious to reintegrate into the Atlantic community and seized upon environmental protection as a possible means. Drafting the convention would also quell criticism from its emerging Green Party that the Bonn government was not doing enough for conservation. The IUCN's Francoise Burhenne-Guilmin recalled that Ministry of Food, Agriculture and Forestry officials circulated a draft prepared by her Law Programme centered in Bonn and then prepared an official text that reflected an effort to create a new type of flexible "umbrella" convention that would coordinate actions for migratory species, and under which various nations could negotiate about how to manage specific animal populations. Final negotiations in the village of Bad Godesberg near Bonn in June 1979 resulted in the Convention on the Conservation of Migratory Species of Wild Animals. It became one of fourteen regional agreements promulgated by the United Nations Environmental Program (UNEP) and makes up one of six conventions that deal with biodiversity under the Convention on Biological Diversity (CBD), which in turn derived directly from the 1992 Earth Summit in Rio de Janeiro. The Rio conventions are "intrinsically linked, operating in the same ecosystems and addressing interdependent issues."[5] This story of the development of the Convention on Migratory Species rehearses the process by which agreements subsequently have developed for the protection of the albatross and fish. The albatross and fish conservation webs are equally complex in their overlapping stages of evolution and in the number of components, as well as in the challenges to implementation and enforcement presented by the collaboration of various levels of governance over multiple, overlapping geographical areas.

Initially a small convention consisting of 15 nations, the Convention on Migratory Species now has 114 members (as of August 1, 2010), adding 10 or so nations annually through the late 1990s. Provisions are binding on members, and as a framework convention, it codifies principles that various regionally focused agreements, such as the 2001 Agreement on the Conservation of Albatrosses and Petrels, may apply.[6]

In providing this international legal framework for animal species that migrate beyond the borders of a single nation-state, the Convention on Migratory Species has played an important role in albatross and fish protection. It attracts the interest of nations with sizable exclusive economic zones whose citizens are likely to be interested in and have an effect on what happens to fish, seabirds, and other migratory animals. Curiously, however, Australia and New Zealand sided with the United States, the USSR, and Japan in taking the minority position of opposing the inclusion of marine species in final discussions about the convention's formulation. Reportedly, their position reflected a need to accommodate the interests of their commercial fishers and their concern about the effects on their sovereignty of negotiations at the time over what grew into the 1982 UN Law of the Sea. Their opposition also reflected their desire to protect the integrity of the Antarctic Treaty System, about which they have always been extremely sensitive, which treats animals on a broad regional basis. Additionally, in the mid-1970s when disagreements about the Convention on Migratory Species were still unresolved, Australia was actively committed to negotiations leading to a regional fisheries management organization in the Southern Ocean, which became the Convention for the Conservation of Antarctic Marine Living Resources (CCAMLR). As a result, Australia took an exclusionary position regarding what areas and species the Convention on Migratory Species should cover. However, a united statement from Africa's delegates insisted that members take a holistic approach for the new convention, arguing that migratory animals "are a common heritage of mankind."[7] This saved the convention, which ultimately included all animal groups and geographical areas within its jurisdiction. However, the two Southern Hemisphere albatross states of Australia and New Zealand did not sign until more than a decade after the treaty's implementation, and Japan has not yet joined, likely because of its desire to continue harvesting whales for "research," which is allowed under the International Whaling Convention. The United States is a nonparty but participating member of the Convention on Migratory Species, but only with respect to marine turtles in the Indian Ocean, as is the Russian Federation with respect to its Siberian Crane.

CHAPTER 8

Globalization

A NUMBER OF ADDITIONAL multinational agreements and actions, not always specifically related to birds but nevertheless foundational elements for our story, happened between the end of World War II and the late 1970s, as typified by the Convention on Migratory Species. They focused on environmental issues, notably endangered species, as concerns about threatened and rare animals developed a strong enough voice to be included in policy discussions, legislation, and international agreements. Changes happened on several fronts. First, national governments reassessed their offshore possessions in light of the postwar global politics that soured East-West relations and led to the Cold War. Second, as the demand for fish protein soared and exposed the high seas to fleets better equipped than ever before, coastal states began to exercise sovereignty over their home and island waters and to monitor foreign vessels around their shores. Third, recognition that unregulated access to the high seas spelled likely doom for targeted animals, such as the great whales, led to demands for better science, with respect not merely to a single animal group or species, but also to the biophysical processes in marine and polar ecosystems. The fledgling United Nations was a suitable and timely instrument for developing a measured response to the threats of unlimited animal harvesting because it was positioned to sponsor and negotiate international agreements about shared resources. That body's Food and Agriculture Organization (FAO) was founded in 1945 in part on the concept of the oceans as a global commons capable of supplying high-quality food worldwide through fisheries management. One of the FAO's most useful functions has been in gathering long-term data from all areas within its purview.

Although the fisheries data that member nations report to FAO are often limited and seriously flawed, FAO is often the only recipient as well as source for such global data. Independent scientists sometimes pass their life's work developing their own data as a check and correction on the data supplied to or generated by FAO, as biologists Ransom Myers and Boris Worm at Dalhousie University in Nova Scotia have done in their long-term assessment of global fish stocks. This is invaluable work because it allows FAO personnel and users of FAO data to identify flaws in data-gathering procedures and results, determine the causes and sources of such flaws, and improve coverage and overall accuracy.

The post–World War II convergence of national values and heightened awareness of limited environmental resources provided the political and economic impetus for countries to reexamine offshore possessions. This included studying flora and fauna previously exploited through unbridled extraction or completely ignored, as with seabirds, because they were thought of no value. For the albatross, this resulted in systematic life history study and research in seabird colonies, led by the United Kingdom, Australia, New Zealand, and France and carried out by scientists committed to Antarctic and sub-Antarctic topics in newly constituted governmental and nongovernmental environmental agencies.

With respect to Britain, a covert Royal Navy operation code-named Operation Tabarin established secret Allied bases in the Antarctic during World War II. Named after a famous Parisian nightclub, this cloak-and-dagger operation was launched in 1943 to prevent Axis vessels from refueling on South Atlantic islands, but also to thwart claims Argentina and Chile had made over the Falkland Islands and neighboring areas. The British government garrisoned soldiers on the Falklands in 1939, along with biologists and earth scientists. Base personnel also monitored enemy vessels and warded off landings on these contested possessions. Another reported but less plausible objective for Tabarin was to draw German scrutiny away from the Allied landings in Normandy. Whatever the reason, through their surveys, the scientists, many already experienced in polar research, reinforced Britain's Colonial Office decision to cede no south polar territories. After World War II, the operation was renamed the Falkland Island Dependencies Survey, which expanded into the British Antarctic Survey in 1962. The military and strategic significance of these initially secret bases reinforced interdependency between politics and science in Antarctic research. The British Antarctic Survey today is based in Cambridge, England, and its scientists have proven to be energetic and long-standing participants in the efforts to study and manage albatross populations. Their long-term studies of the Wanderer colonies on South Georgia provide a benchmark against which population trends continue to be assessed. The news is not good.

Surveying South Georgia's resources in the 1950s greatly enhanced British access to those islands. Prince Philip, Duke of Edinburgh, visited in 1957, toured Grytviken's active whaling station, and paid respects at Sir Ernest Shackleton's grave. He photographed birds, including albatrosses, from the deck of the royal yacht *Britannia* to include in a book he wrote. That same year, the celebration of an International Geophysical Year allowed scientists from sixty-seven nations to coordinate observations of different geophysical phenomena on and around the Antarctic continent.

In 1947, Britain ceded Heard and McDonald Islands (HIMI) to Australia. HIMI, which catches the southern edge of the eastward-flowing circumpolar current, is home to rich fishing grounds as well as foraging albatrosses. Antarctic scientists visit every three years or so to check conditions and take a census of the marine animals. A management plan addresses the area's fragile ocean system, which has been invaded by unscrupulous fishers. Australia also staked a claim to 42 percent of the Antarctic continent in 1947, based on the early expeditionary efforts of immigrant countryman Douglas Mawson. It established an Antarctic Division in 1948, now based in Kingston near Hobart, Tasmania, as an agency of the departments of the environment and foreign affairs. Australia's Mawson research station on Antarctica, the longest continuously run station on the continent, assigned a geologist and biologist to Macquarie Island, Australia's second sub-Antarctic possession, which suffered greatly from sealers' rough handling of its native marine mammals and seabirds from the mid-1800s up to the 1920s. Macquarie Island has been the jumping-off point for expeditions bound for Antarctica. Over the last fifty years the island has proven to be a major study site for long-term wildlife studies, including studies of its nesting albatrosses.[1] Personnel from the Antarctic Division have spearheaded research and animal management on both HIMI and Macquarie Island, including the removal of invasive species from the latter, and have attempted to reduce and eventually stop the death of albatrosses from commercial fishing.

Farther west, on islands in the South Indian Ocean claimed by France, French scientists began long-term studies of important populations of albatrosses in the 1960s. The birds nest on three sets of volcanic islands, each termed an administrative district within TAAF. The largest is the 2,700–square-mile Iles Kerguelen, which Captain James Cook menacingly called Desolation Island. Sealers massively disrupted the indigenous flora and fauna of the main island and islets of this archipelago beginning in 1791, barely twenty years after French navigator Yves-Joseph de Kerguelen-Tremarec had discovered the archipelago that bears his name. France reaffirmed its claim to Kerguelen in 1893 and then renewed its commitment through sponsored expeditions in 1949 and 1950 before setting up a meteorological station for International Geophysical Year studies in 1957–1958. Albatross scientists have worked on Kerguelen and on the other two French archipelagoes almost continuously for fifty years and, as with the British scientists on South Georgia, assembled vital baseline data for monitoring and analyzing the trends and status of key albatross species. After becoming a nature reserve in 1961, the much smaller five-island Iles Crozet have had a permanent French base since 1963. The third albatross-rich district within the French islands includes the small Ile d'Amsterdam and Ile Saint-Paul. Just as on Kerguelen and Crozet, sealers swarmed over these islands after 1789, and settlers introduced livestock such as cattle, pigs, and domestic fowl close to important seabird colonies, including that of the Critically Endangered Amsterdam Albatross, first designated as a separate species by French scientists Pierre Jouventin and J. P. Roux. French presence has been continuous on Ile d'Amsterdam for more than fifty years, more sporadic on crater-shaped Saint Paul, where fires, invasive livestock, cats, and rodents have destroyed native species, although recent eradication efforts have seen some success.[2]

In a secret flag-planting expedition, South Africa annexed Marion Island and Prince Edward Island in late December 1947 and early January 1948, respectively, having recognized the ostensible value of the two islands in the South Indian Ocean for defense and navigation. South Africa's Parliament formalized its claim to the group, which was relinquished by the United Kingdom, and set up a weather station there, periodically approving landings on the inhospitable islands for research. Work on animal biology begun in the mid-1960s has proven important for the five species of nesting albatrosses. Access to the islands, now designated as a Special Nature Reserve, is severely restricted under the national 2003 Protected Areas Act.

Shortly after the end of World War II, a number of coastal nations also began to claim an exclusive economic interest in their national and distant coastal waters by keeping foreign fishing vessels out except under license. In 1947, Chile and Peru claimed sovereignty over their Pacific coastal waters and seabed resources, including marine animals, out to two hundred nautical miles from their shores. In effect, they created exclusive economic zones (EEZs) some thirty years before the 1982 UN Law of the Sea codified the same distance from the coastline as subject to national sovereignty. Five years later, under an international instrument called the Santiago Declaration, Ecuador joined them, thereby bolstering measures to halt foreign vessels from plundering fish stocks in the rich Humboldt Current.

In the late 1970s, Australia, New Zealand, South Africa, Britain, and France also extended sovereignty two hundred nautical miles out from their own coastlines and from coast baselines of their island possessions and dependencies in the Southern Ocean. In 1982, the UN Convention on the Law of the Sea granted coastal nations the right to explore and use, and the responsibility to protect and manage, living and nonliving resources such as fish, natural gas, oil, and minerals within their exclusive economic zones. By mid-decade, half of the world's 142 coastal states had laid claim to two-hundred-mile marine exclusive economic zones. Many proceeded to develop management plans suited to their ecosystems and enforce them with domestic and foreign fishers. According to the Law of the Sea, the host nation establishes laws and regulations regarding fishing in its two-hundred-mile zone for its own and foreign vessels. Furthermore, the nation is responsible for ensuring compliance with international laws and conventions for its own vessels on the high seas. Because more than 90 percent of wild fish are taken from within these national exclusive economic zones, authority over these areas has enormous significance for fish stock sustainability and albatross protection. Declaring exclusive zones has had a long-term effect on the kinds of commercial activities that nations permit within them while also enhancing environmental management in many of them. But because some living resources, such as fish, swim freely back and forth over national political and economic boundaries, nations face the challenge of protecting migratory and boundary-straddling stocks and birds. To conserve and manage such animals, they must negotiate regional and multinational agreements.

Exclusive economic zones also affect what goes on in the neighboring high seas. While fishery authorities claim that the vast majority of wild fish are taken within exclusive zones, making it plausible at least on paper to argue that catch quotas and seabird bycatch issues are likely to be more tractable when under the jurisdiction of a coastal state, enforcement is often absent, uneven, or incomplete. When the 1982 Law of the Sea convention put foreign fishers on notice that they must adhere to a nation's marine regulations, such as they were, many vessels quit fishing in national waters and headed for the high seas, still the "lawless American West" of the oceans. By the mid-1980s, approximately half of the distant-water fishing nations, faced with stricter regulations, quotas, and fees within coastal state zones, hauled their nets and sailed to waters where they could still operate in relative freedom, concentrating on high-value stocks of migratory tuna, for example, that swim through and well beyond exclusive economic zones. When Australia and New Zealand claimed their marine economic zones, the Japanese Southern Bluefin Tuna fleet scaled back and then closed its lucrative operations in their waters, unwilling to tolerate the restrictions and increasing costs of fishing there. The payoff for albatrosses did not happen immediately. Unfortunately, small domestic boats and other foreign vessels operating under national registries replaced foreign fishers and opposed or ignored new regulations designed to reduce bird deaths from fishing. As longline vessel numbers increased within exclusive economic zones, the bycatch of seabirds also increased.

In 1950, at a dinner honoring a leading British geophysicist, National Academies member Lloyd Berkner raised the idea of a new International Polar Year. The gathering's specialists recognized that new technological advances during the recent World War could help answer puzzling geophysical questions in polar regions. British, American, and other players took the idea as an opportunity to make an unparalleled economic survey of south polar areas as well as to showcase their own scientific authority while buttressing sovereignty claims. Passing on the idea to colleagues on atmospheric and geophysical commissions, in 1951 promoters obtained the support of the International Council of Scientific Unions (ICSU), an independent global federation of scientific organizations. The International Council set up a committee to plan the yearlong study and decided on nine programs areas, with Antarctica singled out for special status and fourteen new base camps. The eighteen-month-period from July 1957 through December 1958, designated as the International Geophysical Year and modeled on previous International Polar Years of 1882–1883 and 1932–1933, drew world attention and "big science" toward Antarctica. The International Council organized and received funds for a series of geophysical experiments and drew scientists into collaborative projects. A dozen countries formulated plans for research that dwarfed existing sovereignty rivalries such as that between Britain, Argentina, and Chile. The US and Soviet governments drew up the largest research programs but also called for collaboration. Scientists from several dozen nations coordinated studies of geological, geomagnetic, and oceanographic phenomena employing advanced technology in what turned out to be "an extraordinary moment of Cold War–era cooperation." More than five thousand people from a dozen nations studied Antarctica and manned fifty-five stations altogether, including camps on a number of Southern Ocean islands. This set a precedent for future collaborative global research and international work. Importantly for the albatross, the International Geophysical Year increased international and public interest in Antarctica and the sub-Antarctic.[3]

One product of the International Geophysical Year was the establishment of the Special Committee on Antarctic Research (SCAR). It conducted seismic and gravitational studies at King Edward Point, the old Discovery House on South Georgia, and weather and geological investigations of the Antarctic continent and made plans to continue cooperative scientific studies in these and other Southern Ocean places. It has delivered on this commitment. Meeting every two years, the multinational, nongovernmental SCAR, renamed the Scientific Committee on Antarctic Research in 1961, has provided a platform for developing interdisciplinary collaborative science on Antarctica and the Sub-Antarctic Islands and continues to be an international source of independent advice to the Antarctic Treaty System. Based in Cambridge, UK, SCAR has been actively studying top predators such as albatrosses, their interactions with prey species, and critical sea habitats around Antarctica. During the recent International Polar Year 2007–2008, SCAR-related cruises updated the biogeography of the Southern Ocean, discovering hundreds of new species and defining patterns of distribution, including the possible effects of climate change. For the albatross, the risks include inundation of low-lying nest colonies from sea-level rise; continued expansion of invasive species into subpolar areas due to warming, especially as tourism increases, bringing with it pests that come ashore from ships or even on visitors' boots; withdrawals of birds from nest areas because of heat stress, and the reduced availability of usual foods near breeding islands. For example, a small colony of Light-mantled Albatrosses at latitude 62°S on the South Shetland Islands, almost one thousand miles away from its nearest neighbors, has recently been discovered, ten degrees farther south than was ever documented and the southernmost site for any albatross species.[4]

Work in the Antarctic during the International Geophysical Year provided the momentum and framework for the establishment of a multinational Antarctic Treaty, signed in Washington, DC, in 1959, covering the continent and islands south of 60°S but not marine resources. That would come later, with the development of an Antarctic Treaty System. Seven nations had already claimed areas of Antarctica, some of them overlapping. Avoiding the potential for discord among these claiming nations was an underlying purpose of the Antarctic Treaty. As a result, the Antarctic Treaty prohibits for its duration the assertion of any of these claims as well as the establishment of additional claims, giving an international status to Antarctica as a place of peace rather than discord. The catch, of course, as with all international agreements, is that there is no international body to enforce Antarctic Treaty provisions. Enforcement depends instead on each signatory nation's establishing and enforcing its own laws to support and enforce the treaty. There were twelve original signatories, Argentina, Australia, Belgium, Chile, France, Japan, New Zealand, Norway, South Africa, the USSR, the United Kingdom, and the United States, most of which had already posted claims to areas on the continent. They committed themselves to using the continent for peaceful purposes through research agendas that were collaborative and internationally significant. The treaty currently has forty-five signatories, representing about two-thirds of the world's human population, twenty-seven of which control the decision-making process as consultative parties.

Since entering into force in June 1961, it has been recognized as one of the most successful international agreements ever. Consultative parties meet annually to discuss recommendations and set policy, such as Agreed Measures for the Conservation of Antarctic Fauna and Flora. Antarctic Treaty nations have ratified three complementary treaties: the 1972 Convention for the Conservation of Antarctic Seals (CCAS), the 1991 Protocol on Environmental Protection, and most significant for the albatross, the 1980 Convention on the Conservation of Antarctic Marine Living Resources (CCAMLR).

The acceptance that one has a duty to protect a natural resource in order to hold it in trust for future generations appeared as a part of international law in an international arbitration tribunal in Paris in 1893. Then the question was whether nations had a right to indiscriminately kill fur seals in the North Pacific and the Bering Sea around the Pribilof Islands, or whether the United States, as owner of the breeding islands, had a property right to halt pelagic sealing, which was occurring indiscriminately. The tribunal sided with England and Canada, saying the United States had no property rights over a wild animal on the high seas. But it then arrogated to itself by implication an international legal duty and established a precedent for international law by setting regulations designed to protect fur seals and the fur industry for future generations. Conservationists adopted the strategies drawn up by the tribunal: prohibition of hunting within a specified distance from the islands, seasonal closures, licensing of hunting vessels, reports to governments about catch records, restrictions on type of weapons, and periodic reviews of all regulations. Other conventions began recognizing the interests of future generations and using the tribunal's ruling as a basis for decision making, including the 1946 International Whaling Convention, which stated in its convention preamble that it recognized "the interest of the nations of the world in safeguarding for the future generations the great natural resources represented by the whale stocks."[5]

This concept of intergenerational equity, the second of three environmental principles underlying albatross conservation, is also the founding principle of the World Heritage Trust, formed in 1972. It was elaborated as an ethic by philosopher John Rawls in his *Theory of Justice*, originally published in 1971. One of the strengths of the principle is its applicability to both economic and environmental decision making. Environmental economists have been working to develop various methodologies for quantifying intergenerational equity and factoring it into resource management planning. When ecosystem management and intergenerational equity were joined by the precautionary approach as necessary principles in environmental and economic planning, they formed a sound conceptual structure for more effective decision making.

From the mid-twentieth century on, an increasing number of Southern Ocean nations declared islands and coastal areas as national reserves and marine parks. This practice gradually evolved into a more formal designation for some of these localities as physical or cultural sites worthy of international recognition and protection for future generations. New Zealand, for example, which became a fully independent nation in 1947, as early as 1954 declared Campbell Island a Reserve for the Preservation of Flora and Fauna, and then a decade later named Taiaroa Head, home of Richdale's albatrosses, a Flora and Fauna Reserve. Although Campbell was not the first sub-Antarctic possession to be accorded reserve status and eventually upgraded to World Heritage status, it is the largest island to showcase a costly but amazingly successful experiment to eliminate introduced predators and invasive plants and animals from New Zealand's many islands, particularly those with nesting seabirds.

In 1959, the United Nations Educational, Scientific and Cultural Organization (UNESCO) undertook a successful operation to relocate monuments threatened by the Aswan Dam in Egypt. In 1965, Lady Bird Johnson, wife of then US president Lyndon B. Johnson, held a White House conference to outline a national drive to protect scenic values represented by natural landscapes and the biophysical processes that form them. Three years later, the IUCN made a similar proposal to a World Heritage Trust Conference in Washington, DC, that had begun to stimulate international interest in "the identification, protection and preservation of cultural and natural heritage around the world considered to be of outstanding value to humanity."[6] The IUCN presented its World Heritage initiative to the 1972 United Nations Conference on the Human Environment in Stockholm. With help from the International Council on Monuments and Sites, UNESCO prepared and adopted the Convention Concerning the Protection of the World Cultural and Natural Heritage, known as the World Heritage Convention, which came into operation in 1975.

As of June 2010, 187 states have ratified the World Heritage Convention. It lists 911 properties as harboring outstanding universal value. Included are 704 cultural, 180 natural, and 27 mixed property sites among 151 parties. The World

Heritage concept is so well established that its places are a magnet for international cooperation and receive financial assistance from a variety of sources. Annually, the Heritage Fund makes about four million US dollars available to parties for assistance in identifying, preserving, and promoting World Heritage Sites. By signing the convention, each country pledges to conserve World Heritage Sites and also its own national heritage sites. This means that it will integrate cultural and natural heritage into regional planning to increase public awareness of and tourist activities around a site, thereby adding funds to the local economy. Signatories also agree to supply staff and services to administer comprehensive management plans for their sites and to promote their heritage value among residents and visitors.[7] Currently, nineteen of the albatrosses and petrels protected under the Agreement for the Conservation of Albatrosses and Petrels (see Chapter 22) breed on World Heritage properties, and that list will increase when the Heritage Committee accepts South Africa's nomination of the Prince Edward Islands, with its seabird diversity and abundance, as well as Japan's nomination of its Ogasawara (Bonin) Islands, where Laysans and Black-brows already nest and where the Short-tailed Albatross will join them.

Australia was one of the first countries to ratify the World Heritage Convention. By 2007, Australia had seventeen World Heritage Sites, first the Great Barrier Reef in 1981 and, most recently, the Sydney Opera House in 2007. It worked unsuccessfully to have Antarctica declared a World Heritage Site. Most important for our story, the Heritage Committee designated Macquarie Island for four species, and Heard Island and McDonald Islands (HIMI) for two species, as important albatross nest areas, granting them World Heritage status in 1997.

HIMI, about halfway between Australia and Antarctica, is a World Heritage Site for two reasons. First, although ransacked by sailors, it is the only major sub-Antarctic island group that does not appear to be home to alien animals and plants, making it biologically pristine. Second, Heard Island is home to Australia's highest mountain and only active volcano, the nine-thousand-foot-high Mawson Peak. This adds to the locality's conservation value as possibly the most unstable and fastest growing terrestrial space on earth, set amidst mostly stable, slow-growing, and vulnerable organisms such as corals, sponges, barnacles, and finfish in adjacent waters. Tectonic activity has pushed magma from fissures as much as three miles beneath the crust on Macquarie Island, which is well east of HIMI. These islands are unique exposures of continental drift where the Indo-Australian crustal plate meets the Pacific plate. Much of Australia's remaining exclusive economic zone outside HIMI's marine reserve boundary is closed to fishing except under special license. Macquarie is also an area of outstanding natural beauty and biological integrity, holding about one million pairs of penguins that waddle ashore every spring to commence braying rituals as they court and breed. The Wandering, Black-browed, Grey-headed, and Light-mantled Albatrosses nest on Macquarie's surface. The World Heritage Committee has encouraged Australia to renominate Macquarie Island to be combined with New Zealand's Sub-Antarctic Islands into a single heritage system, and Australia has expressed willingness to do so.

New Zealand claims three World Heritage Sites and has six others under review. One of the three is a set of five island groups that make up the albatross-rich New Zealand Sub-Antarctic Islands, designated as a World Heritage Site in 1998. The Heritage Committee commended New Zealand for submitting a model nomination and suggested cooperation with the secretariat of the Convention for the Conservation of Antarctic Marine Living Resources in order to bolster protection over the marine environment. The islands have high levels of biodiversity and biological productivity, with large animal and plant populations, including 126 bird species, 40 of which are seabirds. Five of the seabird species nest nowhere else and help justify inclusion into a heritage listing. Twelve of the twenty-two species of albatross nest on New Zealand, mostly on these sites. Each island is a nature reserve and has its own management plan. The New Zealand government is working to complete the removal of alien plants and animals from all five Sub-Antarctic Islands.

The United Kingdom's volcanic Gough and Inaccessible Islands in the South Atlantic Ocean received World Heritage designation in 1995. Gough Island has several introduced plants from the sealing era, plus the invasive house mouse. Cool and temperate, Gough Island is a breeding place for one of the world's largest congregations of seabirds, in millions, including penguins, petrels, shearwaters, and a subspecies of Wandering Albatross renamed a full species, the Tristan Albatross. Fifty-four terrestrial and marine bird species have been recorded on the island. There are two endemic land birds and a dozen endemic plants. Additional invertebrates exist on the small volcanic upwelling called Inaccessible Island. Information about mice attacking and eating live Tristan chicks and one or two other birds startled bird conservationists a few years ago, after an ornithologist living on Gough, Geoff Hilton, observed a mouse gnawing on a live chick, "like a cat attacking a hippopotamus."[8] The mismatch works, for fewer than half of Gough's Tristan Albatross chicks, which have no idea how to cope with this introduced rodent, survive to fledging age. In a study for the Royal Society for the Protection of Birds, John Parkes, a Kiwi-based pest-eradication consultant, estimated that it would require 100 tons of Brodifacoum baits, mostly airdropped in winter to the hungry mice, to poison the rodents. Secondary fol-low-up poisoning would require "capturing and holding a proportion of each [bird] population in safety"⁹ until the chemicals had done their work. The entire program would use 750 tons of poison and cost about $2 million.

South Africa ratified the World Heritage Convention in 1997 and nominated the Prince Edward Islands for consideration by both the Ramsar Convention on Wetlands and the World Heritage Committee. In July 2009, the two islands ap-peared on a tentative list of heritage sites on the basis of their outstanding natu-ral beauty, biological diversity, and earth processes. South Africa controls ac-cess to the exposed volcano peaks that are part of the West Indian Ocean Ridge, rising at least ten thousand feet from the ocean abyss. Only researchers brave harsh conditions on these treeless, windswept spaces. The proposed heritage site holds an unusually high percentage of the world population of breeding sea-birds. Twenty-nine bird species, including five albatrosses, nest on the islands. Recent estimates suggest albatross populations there are more or less stable.

The World Heritage status conferred on these five major cold-water alba-tross nesting archipelagoes can be a double-edged sword. On the one hand, de-spite the remoteness of the islands, their special characteristics attract increas-ing numbers of tourists, which, as in the case of the Waved Albatross on the Galápagos Islands, a relatively warm-water site, means extra disturbance and pollution. On the other hand, a World Heritage designation and the individual management plan that accompanies it implies that governments will protect the island ecosystems for future generations and ongoing research. Like a World Heritage Site designation, site designation under the Ramsar Convention, aimed at the conservation and wise use of wetlands of international importance, adds additional layers of protection, management, and international recognition and cooperation, at least in principle, for albatross colonies. South Africa's Prince Edward Islands received the first sub-Antarctic designation under the Ramsar Convention, quickly followed in 2008 by France's thirty-fifth Ramsar designa-tion, consisting of the vast, sea girt habitat around its bitterly cold Kerguelen and Crozet Islands and balmier Amsterdam and St. Paul, all refuges for nesting albatrosses.

Concerns about destructive effects of the international trade in wild plants and animals and their body parts surfaced at an IUCN meeting in 1963 and a decade later at the 1972 Stockholm Conference on the Environment. As a result, twenty-one parties signed a Convention on International Trade in Endangered Species of Wild Fauna and Flora (CITES) in Washington, DC, which went into effect in July 1975 to tackle the uncontrolled trade in such species. If CITES were also to ban trade of endangered wild fish and birds, it would have a significant effect on the health of these species, but the convention has not yet been willing or able to take such protective action. Although parties are bound by convention regulations, national laws have priority, as always. In that sense, CITES is a voluntary convention that frames the issues of trade in wild animals and plants in order to channel domestic laws toward compliance. However, it has tended to focus on exotic species in which trade is already largely illegal, as it is for such well-known or iconic animals as African and Asian spotted cats, crocodiles, and chimpanzees and the body parts of elephants, sharks, and rhinoceroses. CITES has not yet been willing to become actively involved in preventing the endangerment of the ocean's largest commercially traded fish, some of which are threatened with extinction, like Bluefin Tuna and Patagonian Toothfish in the Southern Ocean. The fate of the albatross is linked to the fate of these two species specifically, as we shall see. But when CITES took effect, there was little awareness of such threats to fish or to albatrosses.

Consider the Northern or Atlantic Bluefin Tuna, already listed on the IUCN Red List as Critically Endangered. Environmental writer Charles Clover and author of the 2006 award-winning *End of the Line*, calls it a textbook example of bad trade, "disgracefully overfished on both sides of the Atlantic and in the Mediterranean."[10] He argues that its trade should be regulated under CITES, a step suggested by Sweden in 1992 but blocked by lobbying from the United States and Japan. In 2002, Australia proposed that both the Patagonian Toothfish and Antarctic Toothfish be listed in the CITES Appendix II, which includes species that may become threatened unless trade is strictly regulated. In an analysis of the Patagonian Toothfish population and the CITES (non)listing, Anna Willock, writing in *Uncharted Waters*, suggested that as corals had been admitted under CITES, then finfish could follow. She recommended that parties adopt "Catch Documents for the purposes of CITES permit and certificate provisions under an Appendix II listing."[11] But parties argued that the Convention on the Conservation of Antarctic Marine Living Resources was the body most competent to conserve fish species. In the conference of parties held in Santiago, Chile, in 2002, members endorsed cooperation with CCAMLR and agreed to share information regarding illegal trade in these fish species. But passing the buck like that was the equivalent of a CITES permit to fish illegally for these high-value endangered species. Eighteen years onward, the same issue continues to be debated. In 2010, the United States supported a ban on harvesting Atlantic Bluefin Tuna, but Japan, the major consumer nation of Bluefin, opposed it. A year earlier, Monaco advocated placing Atlantic Bluefin in CITES Appendix I, essentially restoring the magnificent long-distance sea racer to the status of wild animal. Opponents of the recommendation won the day, but calls have grown louder for progress on this matter so CCAMLR and CITES, standing together, can enlist the support of consumers. In 2008, Benjamin K. Sovacool and Kelly E. Siman-Sovacool suggested a two-pronged approach with "concerted action that both reduces the demand for illegally caught animals and increases the penalties and fines for those caught conducting the illegal activity."[12] As we shall see, the consumer part of this strategy is in place, but CITES is still not participating in the effort to save endangered fish species as of the writing of this book.

CHAPTER 9

Commons

FISHING AND THE FREEDOM of navigation on a first-come, first-served basis date back to Roman law, which considered the oceans as held in common, *res communes*, at least conceptually. The attempts of kings and churchmen to claim huge areas of sea notwithstanding, the concept of open access to the oceans, or *mare liberum*, grew into codified practice under the pen of Dutch lawyer Hugo de Groot (1583–1645). He divided the sea into two jurisdictions still recognized today as national territorial waters and the high seas, recodified by the 1982 UN Law of the Sea. In the 1600s, the Dutch proposed extending a country's sovereignty to the distance made by a cannonball shot from the shore. Water beyond the range of ordnance was not subject to appropriation by persons or states but open for navigation, resource extraction, and other lawful uses. This in effect meant that a nation could claim what it could control militarily. Other European nations adopted this position, treating the matter of inshore or territorial seas as a security issue rather than a means of excluding others from fishing.

Great Britain set the standard by defining one marine league, three nautical miles, as the limit of national jurisdiction, the extreme range of a big gun of the time. The United States adopted this measure in 1793, and eventually most European nations followed suit. But differences over territorial jurisdiction remained, and after the end of World War I, the League of Nations brought up the question again. The focus was on rights of access to territorial waters and the high seas, with little or no thought given to the responsibilities these rights entailed. Free access to the more distant high seas evolved into a recognition of the freedom to fish, navigate, and, later, lay underwater cables and pass overhead. Since the 1970s, however, concerns about collapsed and threatened fish stocks have brought the question of shared responsibilities for territorial waters and the high seas into sharp relief.

Exploitation of the world's fisheries increased rapidly after World War II.[1] Postwar prosperity, improved and nationally subsidized fishing fleets, greater fuel efficiency, new fish-finding sonar, lighter synthetic nets, better processing and preservation through onboard canning and flash-freezing, and a rapidly growing world market for wild fish made this expansion possible. But by maximizing takes, the industry fairly quickly encountered stock depletion, as well as the possibility of total collapse of several of the targeted species.[2] Despite this reality, overfishing continues, driven by ever more efficient technologies, the rapidly growing high-end market for fish, and subsidized overcapacity by as much as 27 percent among large nations' commercial fleets. This is rational behavior for a market that operates without an intergenerational equity discount rate, without ecosystem management, and without recourse to the precautionary principle in establishing fishing quotas.

At the beginning of this present century, at least twenty-three thousand vessels larger than one hundred tons, many fitted with sonar screens, satellite phones, GPS plotters, sea-and-climate mapping software, and other sophisticated tools and accompanied by spotter planes and helicopters, were capable of finding and tracking schools of valuable fish hundreds of feet beneath the surface. Millions of smaller craft joined them in this quest. It is the effectiveness of this investment in space-age technology that alarms Clover:

> As a method of mass destruction, fishing with modern technology is the most destructive activity on Earth. It is no exaggeration to say that overfishing is changing the world . . . Overfishing, as a direct result of the demand by consumers in the world's wealthier countries, threatens to deprive developing countries of food . . . and looks set to rob tomorrow's generations of healthy food supplies so that companies can maintain profitability today.[3]

Fish are among the world's most important natural resources. They supply about 15 percent of the protein we collectively consume. The annual rate of wild marine fish taken worldwide rocketed beginning in the 1950s, reached eighty million tons in the 1980s, and peaked at almost eighty-eight million tons in 2000, with a trade value of US$55 billion, a sum that exceeded the gross domestic product of more than 70 percent of the world's nations. Although the Southern Ocean in its peak year accounted for only .01 percent of those tons, it is primarily Southern Ocean fishing that is destroying the albatross family. The Southern Ocean is also home to the endangered fisheries of Southern Bluefin Tuna and the Patagonian Toothfish, both such high-value fish that illegal fishing and marketing of them are strong and growing, more so as stocks become more and more depleted, following the logic of a market absent a factor of intergenerational equity.

In 2002, the FAO estimated that about 47 percent of all wild fish stocks were at full capacity, 18 percent suffered from overfishing, and 10 percent were depleted. But FAO catch figures until recently have been significantly overstated, reporting steady increases in global catch since the 1960s and 104 million tons in 2004, when catches had actually been declining since 1988. The figures were so high, it turns out, because the compensation and advancement of Chinese fishery officials were tied to the size of catches they reported to the FAO.[4] Reginald Watson and Daniel Pauly, from the University of British Columbia, worked for twelve years to explain and correct FAO's overstatement. They estimated that instead of rising since 1988, catches had been declining by about 770,000 tons per year since that time. At the same time, high-seas fishing increased by approximately 70 percent from 1980 to 2000, largely because commercial fishers wanted to escape the regulations nations began imposing in their territorial waters. The largest catch increase was in tuna, followed by deepwater fish like Patagonian Toothfish and Orange Roughy. Both tuna and Patagonian Toothfish are endangered species.[5] Scientists Ransom A. Myers and Boris Worm and colleagues at Dalhousie University in Halifax, Nova Scotia, reported, first in 2003 and then in 2007 with additional data and analysis, that extrapolation of current fish population trends indicates that with no change in industrial fishing practices, a total collapse of the world's populations of larger predatory fish could occur as early as 2048, with cascading effects on the marine and coastal ecosystems. The researchers estimate that the biomass of large predatory fish—such as swordfish, marlin, the larger types of tuna, and other staples such as cod—has been reduced by 90 percent from its preindustrial levels, leaving a precarious 10 percent to be fished today. In regard to development and expansion of commercial longlining in the 1950s and 1960s, they note that

> high abundances of tuna and billfish were always found at the periphery of the fished area. Most newly fished areas showed very high catch rates, but declined to low levels after a few years. As a result, all areas now sustain low catch rates, and some formerly productive areas have been abandoned. In shelf communities, we observed declines of similar magnitude as in the open ocean. The Gulf of Thailand, for example, lost 60% of large finfish, sharks

and skates during the first 5 years of industrialized trawl fishing. The highest initial rate of decline was seen in South Georgia, which has a narrow shelf area that was effectively fished down during the first 2 years of exploitation.[6]

This does not mean that the total biomass of the sea has fallen off. Biological production of life in the sea is at about the same tonnage. It is the tonnage of palatable fish, the authors argue, that has fallen off dramatically, changing the relative balance among all ocean species. Essentially, the older, largest predatory fish are gone.[7]

This interruption in the food chain resulting from the loss of large predatory stocks is having cascading effects on the ecosystems of the world's oceans. And as the more desirable but increasingly rare fish have increased dramatically in price, a nutritional shift of fish protein has occurred and is still occurring, from poorer to wealthier nations. Market pressures and inequalities between First and Third World nations have fostered a form of neocolonialism, whereby wealthier nations have been able to take advantage of the need developing nations have for capital at the expense of these nations' marine natural resources and ecosystems on which their people depend for food and livelihood. Until recently, because of a paucity of expertise, technology, and enforcement authority over foreign vessels within their waters, developing nations have not been in a position to manage their own fish stocks. Wealthier nations have exploited this weakness, some of them while at the same time voicing concern about global environmental stability.

Most destructively, the global fishing industry and the strong world market for fish have encouraged the growth of a flourishing shadow industry of illegal, unreported, and unregulated operators, referred to as IUU fishing and fishers, whose catch tops an estimated US$15 billion annually. "The IUU fish harvest is an unknown percentage of an ill-defined resource,"[8] according to participants at the Organization for Economic Cooperation and Development's 2003 Round Table on Sustainable Development. In some fisheries, illegal, unreported, or unregulated fishing is estimated to take as many fish as the agreed-upon quota for legal fish, effectively doubling the harvest of that managed stock and the effect on its already endangered populations.

The willingness of certain nations to flag vessels with little or no concern for oversight and enforcement of regulations—even the regulations of international conventions to which those nations belong—and of fishing vessels to reflag to such nations, as well as the ability to relocate businesses and capital in the global economy, have facilitated IUU fishing and made bird killing and fish stock destruction much harder to stop. It costs only a few hundred dollars to register with a convenience-flagging country. In 2000, Belize, Honduras, Panama, and St. Vincent and the Grenadines convenience-flagged about 75 percent of the vessels flying flags of convenience. Bolivia appears in the top fourteen convenience-flagging countries, which even include landlocked Mongolia. As of July 2005, 15 percent of the world's fleet of large fishing vessels, including one thousand vessels in the largest category, flew flags of convenience.[9]

Fishery management bodies furnish the best available estimates for illegal fishing and also the greatest potential for doing something about it. Estimates of illegal catches range from 18 percent of the legal quota for Atlantic Bluefin Tuna and 20 percent for redfish worldwide to as high as 50 and even 80 percent for the Patagonian Toothfish in the Southern Ocean and adjacent waters, home for most of the albatrosses. One way to estimate the size of the illegal problem is to calculate the difference between legally reported catches for a specific species and the global trade figures for that species. For example, in 1999 and 2000, the United States and Japan were able to maintain their total imports of Patagonian Toothfish despite a reported twelve-thousand-ton decline in the official catch. Fishers were either storing stocks, waiting for better prices; selling toothfish with doctored documentation; or fishing illegally and smuggling the catches to market. An additional little-known but horrifying problem with illegal fishing is the extent to which boats from some countries force men to work in conditions of extreme hardship and starvation and then throw them overboard as they die from malnutrition, accidents, and exposure.[10]

Illegal vessels offload their catches in ports that Greenpeace and other interested conservation groups identify as providing logistical convenience. Harbors in Mauritius, Namibia, Indonesia, Hong Kong, and Singapore have reportedly accepted illegal toothfish catches. Illegally operating boats switch among these nations as one or the other tightens oversight. Fisheries officials and industry representatives believe that illegal fishing is an increasingly sophisticated operation worldwide that launders money through a shell game of changing vessel names, colors, and flag states and the creation of spurious companies. Some suggest that illegal companies or owners allow one of their older vessels to be intercepted by fishery authorities and led into port to divert attention from other boats as they freely continue to ply their illegal trade. By dint of self–interest, strong persuasion, and leadership, the legal industry is increasingly committed to managing stock sustainably and reducing bycatch. What incentive does an illegal vessel have to manage stocks and reduce bycatch?

Commercial fishing meets the conditions for what the late Garrett Hardin calls the "Tragedy of the Commons"[11]—competition for limited resources, maximum exploitation in the shortest time, social inequity and corruption, and, ultimately, destruction of the resource. In the case of the medieval English commons, where Hardin's theorized tragedy did not occur, the commoners agreed to set reasonable self-imposed quotas for the number of livestock each villager could run on the common land without destroying it. Perhaps extending the metaphor of the village commons to the global high seas is counterproductive. It may mislead and obfuscate in the same way the development of regional fishery regulations and international standards and agreements for sustainable fishing may mask actual ongoing conditions of overexploitation, corruption, and illegal fishing. The concept of the commons gives voice to a jurisdictional ideal of shared ownership of and responsibility for the marine resources owned by no one and yet by all. But there is no globally enforceable mechanism to allow us to monitor and enforce the equal-access and self-limiting practices that would enable us to realize this ideal. Without these capabilities, it is not clear that we will be able to avert the ultimate tragedy.

To date, laws and agreements, national and regional efforts, and private and trade-based systems have been inadequate in regulating when, where, and how many fish we take in the ocean commons where albatrosses live. However, recent breakthroughs on the international stage have added backbone to regulations, as have increasing environmental awareness and citizenship worldwide. In respect to bycatch, albatross scientists recognized from the first that fishing traditions are entrenched and resistant to change except for clear and compelling economic reasons. But the industry has grown increasingly concerned about overfishing and illegal fishing and is therefore open to change. Environmental economists are developing a calculus that commercial fishers are beginning to accept for determining present and future values of natural resources. Fortunately, ecological, ethical, and aesthetic values are beginning to coincide with economic values in regard to marine resources and ecosystems. Once the industry is committed to taking strong and effective measures to correct overfishing and illegal fishing, taking measures at the same time to protect albatrosses and other seabirds may be an inseparable part of the new logic. Consumers hold enormous leverage for this shift, a shift that is nothing less than a global culture and sea change. But they have only begun to realize and act on this power. This is the situation worldwide: a threatened global ocean and severely threatened fish and seabird species, accompanied by a changing and evolving mindset among fishers and the public in regard to how to use marine resources. This could be a hopeful situation.

Part Three

———

BIRDS AND FISH

CHAPTER 10

Fish

GLOBALLY, FISH AUTHORITIES assess and monitor about eight hundred commercial species, eighteen of which live in the Southern Ocean. In 1800, an officer on an American sealing vessel reported the first record of a codlike fish, about eighteen inches long, in the Bay of Isles at South Georgia. It was another hundred years before fishers began, as a side venture to lucrative whaling, to target marbled rockcod and icefish. Efforts failed due to poor equipment, long distance to market, spotty catches, and a lack of capital as whaling was playing out.

Harvests of finfish together with krill began in earnest in the 1950s and expanded across the South Atlantic and southern Indian Oceans as northern nations bolstered sovereignty over Southern Hemisphere island waters. Negotiations about the new and probably restrictive UN Law of the Sea, likely to restrict foreign fleet activities in territorial waters, also pushed fishing fleets into the southern seas. It was that uncertainty that sent vessels from the USSR and other Eastern Bloc nations, such as East Germany, Poland, and Bulgaria, as well as Japan-flagged vessels, south. Advanced market economies did not find Southern Ocean fishing to be profitable, but the command economies of the Soviet Union and Eastern Europe were seeking less costly ways for obtaining animal protein than collectivized agriculture. Because of high prices for fish in Japan, Japanese fishers were also able to fish profitably in the Southern Ocean. These incoming fishers raided stocks in one place after another in a boom-and-bust practice. After just two seasons in the early 1970s, some of the newly tapped finfish around South Georgia were commercially extinct, and more recent tests suggest that others are unlikely to ever recover. Initial targets such as rockcod and icefish collapsed by the 1980s. Second-stage targets, such as lanternfish and toothfish, remained. But as stocks of cold-water species played out, fishers turned to Patagonian Toothfish, which is currently certified for sustainable take in only one place in the world, off South Georgia.[1]

Patagonian Toothfish is by far the most important commercial finfish species in the Southern Ocean, marketed to retailers, restaurants, and customers with the more palatable name Chilean Sea Bass because Chilean fishers first harvested it commercially. It also goes under the name of Black Hake and Mero. Excluding krill and also Bluefin Tuna, which is mostly harvested just north of the Southern Ocean's northern boundary, Patagonian Toothfish accounted for 63 percent of total Southern Ocean catch by weight in the first years of this century and is now among the most lucrative fish species on the world market. The fish inhabit deep and dark seamounts from close to Antarctica ranging northward toward Chile and Patagonia. It is not a bass but a fish acclimated to water as cold as two degrees centigrade. A similar Antarctic Toothfish inhabits even more frigid deeps south of the Antarctic Circle. The tender but hearty white flesh of Patagonian Toothfish fetches as much as US$20–$30 per pound on the wholesale market in North America, Europe, and Japan, and it costs at least $30–$50 per dish in upscale restaurants. Exceeding six feet, the largest specimens of this predatory species—usually females—weigh in at two hundred pounds. They are taken on demersal longlines set near the seabed in waters up to thirteen thousand feet deep. Toothfish mature slowly and live upward of forty years, like albatrosses. By the mid-1990s, Patagonian Toothfish longliners worked stretches of sub-Antarctic waters from Chile east over the Falkland Islands plateau, past South Georgia, and over the Kerguelen plateau as far as Australia's HIMI. Vessels gathered on seamounts and island shelves and also discovered stocks around South Africa's albatross nest islands.

From 1996 to 1999, estimates of total annual Patagonian Toothfish landings, both legal and illegal, in the Southern Ocean topped ninety thousand tons valued at US$500 million. Landings were more than double the tonnage legally permitted by CCAMLR, the fishery management body for the Southern Ocean. Moreover, legal and illegal kinds of longlining during this period killed from 160,000 to 400,000 seabirds within CCAMLR's area alone.[2]

In 2004, the Washington-based National Environmental Trust reported that about twenty, a little more than a third, of the fifty-six nations involved in catching and trading Patagonian Toothfish exempted themselves from CCAMLR regulations. It noted also that elaborate systems existed for smuggling toothfish into key markets such as the United States. Smuggling involves deliberately mis-labeling catches, falsifying permit information, disguising the type and quanti-ties of fish being shipped to market, and offloading toothfish in mid-ocean in order to camouflage their provenance. Vessels unload in ports with minimal or corrupt security, such as Port Louis, Mauritius, a noted illegal fishing haven until its government recently agreed to abide by catch and landing requirements. The two major importers of Patagonian Toothfish are Japan and the United States, followed by Canada and the European Union. The Marine Stewardship Council certifies only the Patagonian Toothfish fishery off South Georgia as managed for sustainability, but absence of certification for sustainability does nothing to stop illegal operators from catching Patagonian Toothfish wherever they can find them, pushing the species ever closer to commercial collapse.[3]

Artisan fishers caught tuna and tuna-type species in tropical and subtropi-cal inshore waters for centuries. Once canning opened up long-distance ship-ments in the early 1900s, demands for tuna increased, especially in Japan, and boosted tuna fishing to the extensive global industry that exists today, despite the threatened status of some species. After World War II, the US government restricted Japan to its own territorial waters, but when the restriction was lifted, Japanese longline vessels headed east and south. A subsidized pelagic fishing fleet operated in Australian waters for almost twenty years under a 1979 bilateral agreement. It was personnel from the Australian Fisheries Division that worked as observers aboard a few of the sixty Japanese vessels that operated off Tas-mania and New South Wales in the early 1990s, yielding breakthrough data on the causes of seabird deaths. After the Japanese fishers quit for less restricted domains when Australia claimed its exclusive marine economic zone, domestic tuna boats replaced them, and hook numbers rose exponentially, topping 17.5 million in 2001.

Although tuna accounts for merely 9 percent of the global trade in seafood, tuna flesh commands the highest seafood prices overall. The quality of the flesh determines the use and market value of each species. There are fifteen pelagic species of tuna, six of which are not principal market species. Canning tunas, low priced and caught in large quantities, include Skipjack, Yellowfin, Long-tail, and Albacore. Bigeye Tuna and Bluefin Tuna fetch the highest prices on the sushi and sashimi market, and Atlantic, Pacific, and Southern Bluefin Tuna are the most highly valued for sashimi. Their raw flesh commands US$30–$40 per pound in Tokyo's Tsukiji Fish Market, Japan's main seafood clearinghouse, and in hip sushi bars, the price for Bluefin is astronomical. The most expen-sive American restaurant of any kind, according to writer Trevor Corson, is a sushi bar in New York City. To celebrate the first business day in the twenty-first century, one customer in Tokyo's Tsukiji Market paid US$400 per pound for a single first-rate fresh specimen, a total of $173,000 for the entire 445-pound fish. In early 2010, a monster 513-pound Bluefin caught off the coast of Japan fetched a record 16.3 million yen, or US$177,733, paid by three sushi restaurateurs despite a downturn in the economy that had pushed Japan into the worst recession since World War II.[4] Bluefin Tuna receives the highest values for its underbelly fat and its steaks, especially in Japan, North America, and Europe. Journalist Sasha Issenberg's *Sushi Economy: Globalization and the Making of a Modern Delicacy* relies on sushi to relay an economic and cultural history of globalization.

The boom associated with post–World War II reconstruction resulted in record tuna catches. Japanese fishers harvested almost 50 percent of the world catch and made sure the nation achieved major standing in the global trade in seafood products. Firms in Japan also exported canned tuna, mostly Yellowfin. In the 1960s, having copied fishing techniques and freezing methods, Asian and European counterparts joined in the tuna binge, and Japanese fishers switched from canning to freezing big specimens of Bluefin and Bigeye for sashimi. It was purse seining for these fish that generated unwelcome publicity because of the large numbers of dolphin bycatch casualties. As longlining replaced purse seining, and as operations shifted to the central and western Pacific and then into the Southern Hemisphere, the amount of dolphin bycatch decreased, but quantities of other bycatch, including seabirds, increased. Many industrial fishers from Korea and Taiwan also switched from purse seining to longlining, compounding the problem for seabirds.

Today, Japan cannot satisfy its demand for seafood and depends on imports to cover shortfalls. The Japanese eat an average of 128 pounds of seafood per person per year, but some Europeans are catching up. Spaniards, for example, consume 97 pounds, compared with 52 pounds eaten by Canadians, 44 pounds by UK residents, and 17 pounds by Americans. EU nationals increasingly desire seafood, including tuna.

With declining populations of the extremely valuable wild tuna, longliners have been pursuing the older, more mature specimens intensively, heading into the farthest reaches of the world's oceans. This depletion of older stockfish, as with albatrosses, has resulted in a threat to the entire species. Almost five million tons of tuna were harvested in 2000, about five times more than in 1950. As Southern Bluefin stocks have taken a hammering, fishers have shifted into the Indian Ocean, where catches have increased over the last decade. This shift means that tens of millions of baited hooks are slung into new waters as fishers quit old ones, giving some seabirds respite while introducing the threat to other seabird populations. At least twenty nations have vessels engaged in tuna longlining, and gear is now being deployed for tuna in all the world's oceans. In the North Atlantic and Mediterranean Sea, fishers are still netting the premium sashimi Atlantic Bluefin Tuna, and they are also caging immature stock for fattening, dangerously reducing juvenile recruitment for breeding. At the start of a five-day meeting in Kobe, Japan, in 2007 of the world's five major tuna fishery management organizations representing seventy-seven countries, the World Wildlife Fund reported that "tuna are fast disappearing, with important stocks at high risk of commercial extinction due to weak management," and that "Atlantic bluefin [tuna], used for high-end sushi and sashimi, is massively overfished and the spawning stock of southern bluefin in the Indian Ocean is down about 90%."[5] Results from Kobe did not please environmental groups. Patagonian Toothfish and Bluefin Tuna are two of the world's most endangered fish species, and harvesting them poses the greatest threat to the future of the albatross family.

115

CHAPTER 11

Management

THE LONG-TERM IRRATIONALITY of unmanaged commercial fishing was apparent to some scientists as early as the 1950s, when UK's Lowestoft Laboratories became a world leader in assessing sea fish populations. Some fishers were concerned even then that Lowestoft director Michael Graham's Great Law of Fishing was correct: a fishery subject to unlimited fishing rapidly became unprofitable. A major development in the late 1960s was the development of regional fishery management bodies that fishers organized to manage one species for maximum sustainable yield through quotas and fish-free zones and seasons. These organizations face extreme challenges, from the incompleteness of hard knowledge about marine ecosystems, to the enormous difficulty of developing accurate fish population models, to nations and fishers uncommitted to sustainability, and to the absence of effective enforcement mechanisms in both territorial waters and on the high seas.

There are currently some forty-three intergovernmental regional fishery management organizations, called RFMOs, worldwide. Four are global in scope, fourteen manage stocks in the Atlantic Ocean, thirteen in the Pacific Ocean, five in the Indian Ocean, one in the Mediterranean and Black Sea, and six in inland waters. More are developing as we write. With few exceptions, their management focuses on a single species of target fish. Their jurisdictions, which sometimes overlap because their target fish stocks inhabit the same waters, vary in size, and they exercise various levels of authority over the fishing vessels of their member nations. Twenty-two have only advisory authority, but twenty have authority to manage stocks, such as tuna (Table 2).

TABLE 2. Regional Fishery Management Organizations with a Management Mandate Involving Albatrosses

ORGANIZATION NAME	YEAR EST.	NO. OF PARTIES	NO. OF ALBATROSS SPECIES PRESENT
CCAMLR: Commission on the Conservation of Antarctic Marine Living Resources	1982	25	12
CCSBP: Convention on the Conservation and Management of Pollock Resources in the Central Bering Sea	1995	6	
CCSBT: Commission for the Conservation of Southern Bluefin Tuna	1994	5	18
GFCM: General Fisheries Commission for the Mediterranean	1952	24	
IATTC: Inter-American Tropical Tuna Commission	1950	14	14
ICCAT: International Commission for the Conservation of Atlantic Tunas	1969	44	11
IOTC: Indian Ocean Tuna Commission	1996	24	18
IPHC: International Pacific Halibut Commission	1923	2	1
IWC: International Whaling Commission	1946	88	
LVFO: Lake Victoria Fisheries Organization	1994	3	
NAFO: Northwest Atlantic Fisheries Organization	1979	12	
NASCO: North Atlantic Salmon Conservation Organization	1983	7	
NEAFC: Northeast Atlantic Fisheries Commission	1982	7	
NPAFC: North Pacific Anadromous Fish Commission	1993	5	
PSC: Pacific Salmon Commission	1985	2	
RECOFI: Regional Commission for Fisheries	2001		

ORGANIZATION NAME	YEAR EST.	NO. OF PARTIES	NO. OF ALBATROSS SPECIES PRESENT
SEAFO: Southeast Atlantic Fisheries Organization	2003	3	12
SIOFA: South Indian Ocean Fisheries Agreement	2006	10	12
SPRFMO: South Pacific Regional Fisheries Management Organisation	2010	open for parties	14
WCPFC: Western and Central Pacific Fisheries Commission	2004	25	18

Source: UN, FAO 2010b.

These twenty management bodies establish catch areas, define seasons and quotas based on the collection and analysis of data, regulate the type of fishing gear and procedures used, and monitor and attempt to regulate the amount of bycatch waste. Some also license and inspect vessels that fish in their waters. The degree to which a fishery management body encourages and facilitates transparency regarding fishing practices and the degree of collaboration with environmental interests depend on its membership. Given the dramatic declines in the sizes and numbers of wild fish and the threatened and endangered status of some of them, fishery management bodies are confronting the inadequacy of their efforts to date and calling into question the data, analytic tools, and management methods they have been using.

Five fishery management bodies with relevance to seriously endangered fish and albatrosses operate in the Southern Hemisphere. The International Commission for the Conservation of Atlantic Tunas (ICCAT), the first of these five bodies to form, ranks third among them in the importance of its area to albatrosses but spans both hemispheres. Signed in Rio de Janeiro in 1966 and put into force in 1969, ICCAT conserves and manages tuna and tunalike fish in the entire Atlantic Ocean and nearby seas, including the Mediterranean.

The 1980 Convention on the Conservation of Antarctic Marine Living Resources (CCAMLR) is unique among fishery management bodies and the most relevant to efforts to save the albatross, but not because albatrosses spend most of their time there—they spend the least percentage of their time in CCAMLR's waters among the five management bodies in whose realms most albatrosses nest and forage. As its name states, it was established to monitor and manage the entire marine ecosystem in its jurisdiction, which is no less than the entire Southern Ocean. It must consider and respond to the pooled concerns of its members about sustainable harvest of all commercial fish stocks while also protecting sustainability of the entire marine ecosystem. Prior to CCAMLR, the International Whaling Commission was the only international body authorized to manage a marine animal in the Southern Ocean. In 1964, the Antarctic Treaty Consultative Parties formulated Agreed Measures for the Conservation of Antarctic Fauna and Flora, limited at that time only to land below 60°S, meaning Antarctica, its islands, and surrounding land shelves. In 1972, the Consultative Parties turned their attention to the ocean environment when the Special Committee on Antarctic Research set up a subcommittee on Living Resources. A group of its specialists expressed concern about the keystone role of krill in the short food chain in sub-Antarctic waters. At the time, Soviet and Eastern Bloc vessels were trawling for finfish and krill, and within a generation had exhausted supplies of Marbled Rockcod and some bottom-dwelling fish. Some parties were concerned with the threat of a potentially unlimited take of krill to the recovery of seal populations as well as to the marine ecosystem. In response to these concerns, the parties unanimously agreed to negotiate a new treaty for Antarctic marine living resources based on the concept of the ecosystem. In early 1978, thirteen delegates established a convention framework in Canberra, Australia. Most pressed for a clear boundary designation for the Antarctic ecosystem as well as a clear statement of the conservation standards that would govern the convention. But Japan and the USSR argued for using the standard language of fisheries agreements in respect to harvesting krill and opposed formulating new conservation standards for the Southern Ocean. Boundaries remained a contentious issue because of distant-nation sovereignty over Southern Ocean islands.

The initiative almost collapsed in the second consultation in Buenos Aires in July 1978, fractured by sovereignty concerns and disagreement over the proposed rule by consensus instead of majority. Sovereignty was a major issue for France, which did not want Kerguelen and Crozet to fall within the new convention's jurisdiction. From CCAMLR's first conception, the boundary had been intended as the Convergence or Polar Frontal Zone, to encompass what is broadly known as the Southern Ocean. The trick was how to map it with rectilinear coordinates that captured the appropriate biological boundaries. It has since become clear, for example, that some adjacent waters that include remaining Patagonian Toothfish stocks should also have been included. Hard bargaining took place, and the parties finally agreed in a late-night session to define the area as south of 60°S but with a legally acceptable moving boundary. Differences over sovereignty persisted. Trusting that no one really intended to torpedo the package, parties allowed all coastal state claimants, including France, to state their positions while they continued to work toward agreement.

CCAMLR came into existence in May 1980 and into force in April 1982 after ratification by eight of the original signatory states (Table 3). Negotiators accommodated France's retention of sovereignty over its exclusive economic zones around sub-Antarctic Kerguelen and Crozet. South Africa claimed the waters around the Prince Edward Islands; Australia those around Heard, McDonald, and Macquarie Islands; and the United Kingdom those around South Georgia and the South Sandwich Islands, all of which fall within the convention's area.[1] These sovereignty claims do not necessarily conflict with conservation objectives for marine living resources. However, CCAMLR's approach and the broader array of shared scientific knowledge it required for decision making challenged the single species–based strategies of some fishing nations. Consensus-based decisions were slow in coming, and once they did, enforcement of CCAMLR standards and regulations depended entirely on the will of its member nations as well as on the good faith and effort of nonmember flagging states. So CCAMLR got off to a slow start. It didn't begin enacting effective measures until the 1990s.

The protection of intergenerational equity is assumed in the convention's mandate to conserve Antarctic living marine resources. In negotiations and decision making, CCAMLR members also rely on the other two basic principles of environmental management: ecosystem management as derived from the Ramsar Convention, and the precautionary approach. From its inception, CCAMLR adopted the precautionary principle as a basis for management, and over time has drawn extra support from UN fishery agreements also based on this management premise in regard to boundary-straddling and highly migratory fish stocks. The precautionary approach calls for states and other management entities to act conservatively, that is, more strongly on the side of the worst-case scenario when information is uncertain, unreliable, or inadequate. But uncertain, unreliable, and inadequate information characterizes fishery management. The precautionary principle does not suggest that decisions be deferred until certainty, or a greater degree of certainty, is achieved. The adoption of the precautionary principle in regard to fisheries and ecosystems means, in effect, that a managing fisheries agency such as CCAMLR is almost always obliged to adopt the more conservative practices and quotas. Although fishery bodies are better positioned than the FAO to research marine issues, they are hard put to explore and understand stock interactions. In accepting responsibility for the marine ecosystem, even though fully acting on it may be out of

TABLE 3. Overview of CCAMLR

CCAMLR was established in Canberra, Australia, on May 20, 1980, and went into force on April 7, 1982, as an instrument of the 1961 Antarctic Treaty (AT). It is also an example of a regional fisheries management organization.

CCAMLR is the cornerstone in the regulatory framework for conserving marine organisms using an ecosystem and precautionary approach through multispecies management in its area of the Southern Ocean.

It has jurisdiction over the living marine resources within the circumpolar area south of 60°S latitude and the Antarctic Convergence zone (FAO statistical areas 45, 58, 88).

As of December 2010, its membership was 25 members (including the EU), and 9 party states. It is open to parties in the AT and to any state or regional economic authority engaged in research and/or harvesting living marine resources in its jurisdiction.

Composition:
- The Commission is the decision-making body that meets annually to discuss management issues, including catch limits on krill and fish. The Commission has two sub-committees:
  - Standing Committee on Administration and Finance
  - Standing Committee on Implementation and Compliance.
- The Scientific Committee, which is a forum for consultation regarding data and research into marine living resources, advises the Commission. The Commission tasks the Scientific Committee to set criteria for conservation measures, assess sub-Antarctic wildlife populations, study the effects of fishing, and formulate plans for research. The Commission must take full account of the advice of the Scientific Committee, which has two Working Groups:
  - Ecosystem Modeling and Management
  - Fish Stock Assessment.

In addition, there is an ad hoc Working Group on Incidental Mortality Arising from Longline Fishing (WG-IMALF, 1993; renamed WG-IMAF, Incidental Mortality Associated with Fishing, in 2001), which answers to the Scientific Committee.

A permanent Secretariat in Hobart, Tasmania, administers CCAMLR.

Any state acceding to CCAMLR is entitled to be a member of the Commission, and all members of the Commission are entitled to be members of the Scientific Committee. The chairman and vice-chairman of the Commission and the Scientific Committee are elected to serve in an honorary capacity and hold office for two years.

121

On matters of substance, namely Conservation Measures—which are binding—
the Commission takes decisions exclusively by consensus, whereas other less important decisions, such as Resolutions, are made by a simple majority of members
present and voting, and are nonbinding.

The Commission and the Scientific Committee are charged with cooperating with
other appropriate agencies, including the UN instruments, the IWC, SCAR, ACAP,
and additional RFMOs.

reach for years to come, CCAMLR has stepped ahead of other fishery bodies.
Despite the weaknesses of a consensus-based rule and some uncommitted and
noncompliant members, CCAMLR has also taken the lead in managing its fish
stocks, reducing bycatch, and taking measures to eliminate illegal fishing from
its waters. The greatest challenge, of course, has been and will continue to be
that of enforcement across the vast surface of the Southern Ocean.

By the close of the 1990s, CCAMLR had passed more than forty conservation
measures and its Commission had adopted two special resolutions. The first,
in 1990, advocated an end to large-scale pelagic driftnet fishing. The second, in
1993, urged members to harvest boundary-straddling fish stocks responsibly,
both within and beyond CCAMLR boundaries. In 1995, as the UN Fish Stocks
Agreement was still being negotiated, Australia circulated the agreement in
draft form to CCAMLR members despite strong objections from Japan. Australian representatives believed it would embolden and strengthen CCAMLR
members and the authority of CCAMLR's Standing Committee on Observation
and Inspection in their efforts to build a more effective monitoring and inspection scheme for treaty waters. One of the greatest values of what became the
Fish Stocks Agreement was signatories' recognition of the need to ensure consistency between international conservation and management measures and
those within areas of national and regional jurisdiction. This was important for
CCAMLR in dealing with longlining for toothfish because under the Fish Stocks
Agreement, coastal CCAMLR members must respect CCAMLR's conservation
measures in their own exclusive economic zones. Although it is still legally possible for a coastal state member to opt out of these measures, for the most part
the five states that claim islands within the CCAMLR area have not diluted its
regulations. In fact, they have tended to adopt even more stringent measures in
their waters. The United Kingdom has been active in the surveillance of South
Georgia's waters, and Australia has patrolled off remote HIMI. In the first case,

the UK's robust stance likely drove illegal operators eastward into the Indian Ocean, suggests John Croxall. In such vast waters, driving illegal fishing out of one area just pushes it into another. And there is concern about how comprehensive Australia's recent HIMI plan is for conserving the delicate marine ecosystem and how forceful its implementation will be. Since the 1995 Fish Stocks Agreement was signed, CCAMLR has increasingly aligned itself with the principles and best practices the agreement calls for. Although the Fish Stocks Agreement goes further in the reach of its standards and measures, which extend from the water to the market, it provides little elaboration of the multispecies aspect of ecosystem management. In this, CCAMLR remains the model.

Two other fishery management bodies with relevance for the albatross, the Indian Ocean Tuna Commission (IOTC) and the Commission for the Conservation of Southern Bluefin Tuna (CCSBT), came into being in the mid-1990s, years that were, in retrospect, a major turning point in this story. These two fishery management bodies, along with the Western and Central Pacific Fisheries Commission (WCPFC) and the International Commission for the Conservation of Atlantic Tunas (ICCAT), have the greatest presence of albatrosses in their waters, and they each have working groups for seabird bycatch issues. The Indian Ocean Tuna Commission was founded in 1993 under the constitution of the UN's Food and Agriculture Organization and became operational in 1996 (see Table 2). It manages sixteen tuna and tunalike species in both the eastern and western sections of the Indian Ocean and surrounding seas, an area that covers waters westward from just east of Tasmania almost to the tip of Africa, from 160° to 35°E longitude. It encompasses most of the Indian Ocean, with the African mainland in the west and Southeast Asia through to Northwest Australia on the east. The southern boundary follows a line of latitude from 45° to 50°S. Membership includes Australia, France, the European Union, and the United Kingdom, and the Commission coordinates research, gathers scientific data on populations, and adopts management measures. Voting is by a two-thirds majority, but any voting member reserves the right to withdraw from a binding measure. Despite the weakness of the option to withdraw, this is at least a more honest procedure than those of consensus-ruled commissions whose members adopt a regulation only to ignore it when it comes to their own vessels.

The 1994 Commission for the Conservation of the Southern Bluefin Tuna (CCSBT) falls within a latitudinal band of 50° to 30°S that almost spans the globe, covering about twenty-eight million square miles of the Southern Hemisphere just along CCAMLR's northern boundary. Eighteen species of albatross breed or forage in its waters. During the peak years from the late 1950s through the mid-1960s, when the harvest topped more than eighty-six thousand tons in 1961, longline fishers harvested 80 percent of all bluefin. As early as the 1980s, when many believed catches had grown unsustainable, the three major tuna fisher nations, Japan, New Zealand, and Australia, began to impose quotas on their own bluefin longliners in order to manage recovery. After working together voluntarily for many years, the three reached a formal agreement in 1994 because of concerns about depleted stocks. Today, catches are officially below one-fifth of the 1960s tonnage and continue to decline. Headquartered in Canberra, Australia, the CCSBT sets catch limits, sponsors research on stocks, coordinates flag-member activities, and cooperates with similar management bodies.

The Western and Central Pacific Fisheries Commission (WCPFC), with seventeen albatross species present in its waters, was not formed until 2004 as a product of the 1995 Fish Stocks Agreement. As a result, it began with a set of standards that put it ahead of the other fishery management bodies and on a level with CCAMLR.

A major challenge to fisheries management was—and still is—to determine what limits will allow optimum catch and at the same time ensure sustainability of a fish population. The Lowestoft Laboratories' Michael Graham had argued early on that if Europe had a reliable way of calculating stock sizes and the effects of fishing pressures, it could avoid developing overcapacity in its fishing fleet. He also believed that fishery managers could use reliable data about stock size to determine the optimal point between fishing intensity and sustainable population size. But complete accuracy in developing and using a fish-stock-population model is impossible—there are too many variables and unknowns. Furthermore, some scientists who have developed accepted models have come to the conclusion, as did Daniel Pauly, "that each generation of fisheries scientists accepts as a baseline the stock size and species composition that occurred at the beginning of their careers."[2] For example, overly optimistic fishery quotas were used for thirty years based on a scientific model that assumed that the number of juvenile fish reaching exploitable size each year was not related to the number of parents in the sea. Other excessive quotas resulted from using a population model based on the assumption that a small fish population, because of lower demand on more resources, would be more productive than a large one. But now scientists know that small fish populations after a certain point become unhealthy populations, precipitated into a downward decline. They are *less* productive because of what is known as the depensation or depopulation effect, and scientists can't distinguish depensation from the effects of fishing pressure in time to correct the error.[3]

As fishery management bodies developed, at first they calculated what they called the maximum sustainable yield (MSY) for the species they managed, using the best models and data they could. But this method was based on the assumption that small populations would by nature increase their size. It took some time for it to become clear that this formula seriously shorted the sustainability side of the relationship. Fishers were taking up to half of the total spawning stock annually and population numbers were falling rapidly. The fishery management bodies then replaced maximum sustainable yield with smaller quotas called total allowable catch (TAC). But these smaller quotas did not take into account illegal fish harvests. Illegal fishing and the fact that some nations consistently underreported catch numbers contributed to a further erosion of stocks to well below their optimal or sustainable size, resulting in the drastic stock depletions worldwide that Myers and Worm reported in 2003. Using the models was clearly counterproductive.[4]

Despite some significant progress toward their objectives and purposes, fishery management bodies, especially those governing endangered Bluefin Tuna and Patagonian Toothfish, still face almost overwhelming challenges:

· Insufficient knowledge
· Fishery management focused on a single species instead of the ecosystem
· Overcapacity and government subsidization of fishing fleets
· Uncertainty and imperfection of fish population modeling and the resulting ineffectiveness of or actual harm from the imposition of catch quotas
· Immense complexity of ecosystems and attempts to manage them responsibly
· Noncompliance with environmental regulations and absence of effective enforcement mechanisms
· Fishers and nations not participating in fishery management agreements and thus not subject to compliance with catch regulations
· Exploitation of developing nations, neocolonialism by wealthier countries
· Illegal fishing and tragedy of the commons
· Flags of convenience sold by nations with little or no intention of overseeing fishing activities of the flagged vessels
· Acceptance in ports of illegal, undocumented landings of fish
· Failure by nations to require legal fishing by their nationals
· Wholesalers, retailers, restaurants, and individual consumers indifferent to the source and legality of the fish they sell and consume

The development of additional tools to manage fish stocks sustainably will do much to assist in the general culture change that is necessary to avoid the tragedy of the commons. Sustainable management of a natural resource like a fish stock requires not only reliable scientific data and effective, highly complex modeling on the species level but also the application of a differently derived discount rate for determining future value from that used in traditional or pre-environmental market analysis. The traditional calculation makes it economically logical to deplete and even exhaust stocks. Many concerned fishers also believe that sustainable management requires trustee ownership of the resources by those who are profiting from them. Ownership takes the form of fish stock rents or rights. In some of the trials of fishery management using a rights system, the fishers are finding that ownership is providing sufficient incentive to the owners not only to limit their harvests but also to pursue and prosecute illegal fishing aggressively in their fishing grounds. The downside of a rights system is that, as with all market enterprises, consolidation in the hands of a few takes place over time. Clover summons the image of fishing rights owners vacationing in Florida while running a secondary market in fishing rights by cell phone.[5]

As we shall see, in significant ways all five of these fishery management bodies still suffer from the weaknesses inherent to all agreements and conventions of nations. But they have been able to learn from models like CCAMLR, benefit from changes in the culture and economics of commercial fishing since 1994, and take more knowledgeable and authoritative measures to ensure sustainable stocks and reduce bycatch. Although their potential has yet to be realized, they are key pieces in the development of an environmental regime favorable for the protection of albatrosses.

CHAPTER 12

Crisis

IN THE 1960S AND '70S, many scientists considered the outlook for albatrosses had brightened since the destruction due to sealing and whaling. General optimism abounded about the results of species conservation and the restoration of nest island habitats. But some scientists working independently in different places, such as French biologists Pierre Jouventin and Henri Weimerskirch; Tasmania-based bird experts Nigel Brothers, Rosemary Gales, and Graham Robertson; and British Antarctic Survey ornithologists such as John Croxall and the late Peter Prince, were becoming concerned about a slight but noticeable downward trend in the sizes of the nest colonies they were monitoring. The reason for this decline was not immediately evident, nor did the scientists have reason to believe it was general. One by one, however, they began to raise the alarm. Analysis revealed that declines probably started several years earlier than first noted but numbers were trending downward on several nest islands far away from one another. Something or someone was killing them.

The South Georgia group was the first to notice and publish their observations about declines. A second team working on French islands in the Indian Ocean followed suit. Pierre Jouventin, a research director at the laboratory of the Centre National de la Recherche Scientifique (CNRS) in Montpellier, specialized in the biology of Antarctic marine animals. Anomalies in the status and numbers of bird populations on France's sub-Antarctic islands captured his attention. In the 1980s, Weimerskirch joined Jouventin to fulfill his French military service by studying the breeding behavior of albatrosses in France's possessions.

Weimerskirch is now a research director at the CNRS laboratory in Chizé in western France. By the mid-1980s, they had published information about the decline of Crozet Wanderers, suggesting for the first time that it was due to fishers working close to nest islands. Over the years, Jouventin and Weimerskirch have collaborated on over thirty publications, mostly on Southern Hemisphere birds and mammals, traveling on a supply ship that makes the ten-day trip from Ile de la Reunion, France's tropical paradise, to the isolated, cold, and jagged spaces of Kerguelen and Crozet.

At the turn of the 1990s, with field support, Jouventin and Weimerskirch collaborated on an important breakthrough using satellite technology. They fitted six male Wandering Albatrosses with transponders and observed their foraging behavior around nest sites on Possession Island in Crozet from January through March 1989. The birds surprised them. Satellite data showed that while its mate incubated their egg, various males tracked between 2,300 and 9,444 miles in a single feeding trip, returning to the nest to trade places after as long as thirty-three days, almost five weeks. Birds headed off in all compass directions, some, as expected, into the icy waters of Antarctica well south of 65°S, while others cruised toward Madagascar, north of 40°S. The prevailing winds influenced foraging strategies. Instead of relying on eyesight, the birds responded to and used wind systems to swing them great distances in tack and zigzag flight to detect and take prey, which they apparently could smell from miles away. Individual birds favored head or side winds, avoiding tail winds and landing on calm days. The scientists' first article about flight tracking appeared in *Nature* and stimulated satellite research that continues today.[1] Their work and that of others using their innovative methods provided the essential data for mapping albatross nesting and foraging patterns and numbers in the 2004 BirdLife International publication *Tracking Ocean Wanderers*. That publication added another critical component to efforts to save the albatross: the ability to conduct overlay analysis of the areas and periods in which commercial fishing and albatrosses converge.

British seabird scientist John Croxall regards Weimerskirch as the most gifted seabird scientist anywhere in the world today, with no comparison. That's not insignificant praise coming from someone many people would also describe in those terms. Weimerskirch has continued to refine tracking systems among seabirds. Wondering how efficient birds proved as fliers, and working with physiologists in California in fitting test birds with lightweight satellite transmitters, activity recorders, and heart-rate monitors, Weimerskirch confirmed that albatrosses as a fixed-wing bird group employ a remarkably efficient mode of movement. His birds showed heart rates merely 10 to 20 percent higher in flight than at rest, which is barely one-tenth the difference for birds that flap their wings to fly. Although the long-winged Wanderers are superbly adapted to the temperate seas, when water temperatures fall below average around nest islands, the birds have to work harder to secure enough food for their chicks. Sea temperatures thus affect breeding and fledging success.[2]

In the 1970s and '80s, working on nesting seabirds on Kerguelen and Crozet, notably the albatrosses, Jouventin and Weimerskirch observed that overall breeding numbers, which had climbed back after being hit during sealing and whaling times, were beginning to sink again. They worried about birds on Crozet, which boasts one of the most diverse seabird communities in the world, with thirty-four species totaling thirty million individuals. By analyzing population data from the previous twenty years, the scientists concluded in the late 1980s that the Wandering Albatross was in steady retreat. Decline had set in twenty years earlier and grown progressively worse. Ominously, adult losses were increasing, which led them to speculate that because of "the increasing number of reports of accidental or deliberate killing by fishermen, and of entanglement in fishing lines and nets,"[3] commercial fishing was a likely cause.

This speculation hardened. Comparing penguin numbers, which had rebounded from commercial slaughter on both Kerguelen and Crozet, with the numbers of albatrosses and petrels, it was clear that although some seabird numbers fluctuated with natural conditions, the albatross trends were clearly down, and in some cases sharply, dropping by 50 percent for Wanderers on three nest islands on Crozet during the 1970s. This worked out to an unsustainable 5 percent decrease annually. Similar observations on Australia's Macquarie Island more than 4,300 miles east of Crozet indicated that its small population of breeding Wanderers, merely twenty-seven pairs, also fell at the same rate in the three seasons after 1975.

Jouventin and Weimerskirch reported that fishing tackle accidentally trapped birds. Initially, this pointed to commercial trawling begun in the late 1950s, then intensifying as Russian vessels fished for cold-water species over the sunken but shallow Kerguelen Ridge frequented by breeding albatrosses and petrels. But the French scientists also wondered about risks posed by more and more vessels that used newer fishing techniques for catching krill. Moreover, Japanese longline tuna fishers who were working warmer waters in the north had recovered and turned in a number of albatross leg bands. Jouventin and Weimerskirch concluded that expanded fishing was killing birds, and their populations "may be unable to compensate for accidental takes by fishing lines and nets."[4] Experienced adult birds were found dead from bill and face wounds.

Others were observed sitting with upchucked hooks around their nests. Gender also seemed to be a factor. Significantly higher numbers of females were failing to come back to their nest colonies. This suggested that while their mates headed south to fish along the edges of Antarctic ice, the females were foraging northward into subtropical waters where tuna longliners worked.[5] Even a small reduction in adult survival was likely to induce a rapidly increasing downward spiral in colony populations. Sealers, whalers, and farmers no longer posed a threat to the albatrosses, but fishers, it appeared, were replacing them. Both scientists expressed fears about the costs of commercial fishing to the fate of the albatross, especially of the Wandering Albatross and the already endangered Amsterdam Albatross, the latter of which Jouventin together with colleague J. P. Roux had described as a new species in 1983.[6] What could be done to stop these losses? they wondered. They also investigated the conservation status of other seabirds that breed on these French sub-Antarctic islands and more recently focused on the eradication of invasive species and the restoration of these lonely habitats for the protection of their native birds and mammals.

While the French were assessing the situation on their islands, John Croxall had also begun to calculate declines in the Wandering Albatross colonies he was monitoring on South Georgia. This Cambridge-based scientist has studied the roles and interactions of mammals and birds in marine ecosystems over a thirty-year span, most specifically in the Southern Hemisphere and Southern Ocean. After completing a doctorate on the ecology of ascidians, filter-feeding sea squirts, Croxall became interested in seabirds oiled by coastal vessels, first as a senior research associate in zoology and then as director of the oiled seabird research unit at the University of Newcastle-upon-Tyne in the early 1970s. After joining the British Antarctic Survey in 1976, he became head of the Birds and Mammals section, the equivalent to holding a university-endowed chair.

Croxall has written and contributed to books and hundreds of articles reporting his research on marine animals, especially albatrosses and penguins, and on Southern Ocean ecosystem measurement and management. He has studied seabird reproduction, population trends, and feeding ecologies in colonies on sub-Antarctic islands, primarily South Georgia, and has investigated the diving behavior of seals, penguins, and petrels. As mentioned earlier, he was one of the three editors of the 1984 International Council for Bird Preservation's *Status and Conservation of the World's Seabirds*.

Croxall was a member of the UK delegation to CCAMLR from 1987 to 2003. He served as president of the British Ornithologists' Union, one of the oldest scientific organizations dealing with birds, from 1994 to 1999, and from 1998

Wandering Albatross, a casualty of industrial fishing with longlines. Courtesy Graham Robertson.

to 2004 he chaired the council of the one-million-member Royal Society for the Protection of Birds. The RSPB supports conservation efforts on an international scale, including vulture recovery in India, crane conservation in China, and albatross protection in the Southern Ocean. In 2000, it launched a Save the Albatross campaign at Britain's annual BirdFair. As council chairman, Croxall helped guide habitat purchase and restoration for British wild birds, promoted their study among youngsters, lobbied the government about wildlife-friendly agricultural practices, and strengthened laws for wild animals—all the while keeping an eye on his albatrosses.

In 1996, Croxall received the Marsh Award from the Zoological Society of London, and in 2004, the British Ornithologists' Union's Godman-Salvin Gold Medal for his distinguished work on seabirds, especially albatrosses. In the same year, Queen Elizabeth named him a Commander of the Civil Order of the British Empire for services to ornithology, and a year later, he was elected a Fellow of the Royal Society for pioneering studies of the global marine system, notably of seabirds.[7] In 2006, he retired as head of Conservation Biology for the British Antarctic Survey and became chair of BirdLife International's Global Seabird Programme. Croxall is a forceful and effective spokesperson for albatrosses and other seabirds. The number of significant organizational and institutional roles he has taken and the honors he has received have given him the global standing to facilitate communications and exchanges of information between national governments, regional and international organizations, nongovernmental organizations, scientists, and high-profile individuals who take an interest in seabird conservation.

Croxall consistently worked to integrate input from British Antarctic Survey research with the research of other members of the National Environment Research Centers, in which the British Antarctic Survey participates, and with similar research in UK universities. It was in 1976 that Croxall joined the British Antarctic Survey on South Georgia as director of its higher predator section. In the seasons of Lance Tickell's studies on Bird Island, the Wanderer population there had rebounded from the effects of sealing to total about 4,600 pairs. In the early 1960s, adult survival on South Georgia exceeded 94 percent, and just over 30 percent of young birds survived to become breeders. But using Tickell's numbers as a baseline, Croxall and colleagues noticed that survival rates for both age cohorts on Bird Island had started to fall, dropping at a rate of close to 1 percent annually, so that by the late 1980s only 1,400 pairs remained on that well-studied site. Currently there are fewer than 1,000 breeders there, and over the past forty years there has been a 50 percent loss in the number of that island's nesting adults.

Croxall first speculated that changing conditions in the Australian wintering grounds, identified by Lance Tickell and Doug Gibson, were hurting the birds. Perhaps, though less likely, their decline was because of increased competition from Grey-headed Albatrosses.[8] He didn't know. However, as coeditor for the *World's Seabirds* book, he found a clue in the details submitted by his French colleagues.[9] He also found more than a clue in what Australian scientist Nigel Brothers was observing onboard vessels of the Japanese tuna longline fishing fleet southeast of Tasmania.

Employed by Tasmania's Department of Parks, Wildlife and Heritage, Nigel Brothers was the first scientist to systematically observe and prove how and to what extent deep-sea fishers accidentally killed seabirds, specifically albatrosses. At the time of his work in the 1980s, many people interested in seabirds judged that adequate protections had been imposed to conserve albatrosses.

Government officials had officially prohibited persecution by declaring many albatross islands as parks or sanctuaries and had begun campaigns to rid the islands of introduced predators. Visitors had begun to land on some nest islands to observe and photograph albatrosses rather than steal eggs or kill chicks. Game and wildlife officers were ensuring that tourists did not disturb nesting birds.

However, by analyzing Japanese fishing data from 1981 to 1986, Brothers estimated that albatrosses were dying as bycatch, that is, as incidental or nontarget catch, in the tuna fishery south of 30°S where Japanese vessels each year played out just over one hundred million baited hooks swinging from longlines. As a scientist, Brothers was also aware of the bad news coming from Crozet, South Georgia, and Tasmania's Macquarie Island. In 1988, Brothers was able, through the systematic observations he and three other Australian Fisheries observers made onboard five Japan-based pelagic tuna longline fishing vessels off the east and south coasts of Tasmania, to establish with empirical certainty the cause of the noted declines in albatross populations in breeding colonies. Standing on the deck, the observers watched the crew clip unweighted branch lines called snoods, up to 115 feet long and each armed with a baited steel hook, onto the monofilament main line. With the ship moving at seven to ten knots, usually in deep ocean waters, the crew clipped a snood to the main line every six seconds or so and chucked it over the stern. Weights along the main line, also calibrated to maximize catch, carried the snood hooks to their predetermined zones, often more than three hundred feet down, where the schools of target fish, in this case tuna, were swimming. The snoods were measured to dangle baited hooks at eye level of the Bluefin Tuna as they swirled through the water. Above them the main line stretched like a garland, hanging its fruits between buoys armed with lights and transponders.

As the crew played out the line over the stern's starboard side, which took about five hours, each of the two to three thousand hooks with its shiny bait of squid or fish tossed in the vessel's wake before sinking on the line that eventually stretched behind the tuna vessel for sixty miles or more. Brothers noticed that propeller turbulence often tangled the branch lines and that incompletely thawed baits tossed about for a longer time before they sank, extending their exposure time to hungry birds. From the deck, the observers saw a fluttering umbrella of seabirds trailing and fanning out five hundred yards or so behind the stern. Every thirty minutes during line deployment the Australian observers recorded sea and weather conditions and every successful or unsuccessful dip a bird made on the sinking baits, including how far an albatross was astern when it swooped down. It was not always possible to tell whether a bird had hooked itself. But a pair of raised and then thrashing wings in the boat's wake identified a snagged individual with a hook embedded in its bill or throat. The panicked albatross tried desperately to lift away, only to succumb to the counterpull of weights designed to carry the line two or three hundred feet below the surface into the zone where tuna cruise, or even deeper for demersal longlining, along the seabed where Patagonian Toothfish swim. Watching the thrashing bird, the crew knew they had lost the bait as well as a potential fish. Unless it broke free, the bird lost its life. Occasionally a bird was able to free itself from the hook before being dragged under, but it likely lost its life later from a bill or throat wound. Although some hooked birds were not seen again, as sharks likely took them, when the crew hauled the line back in, they encountered the hooked birds, emerging as soggy carcasses swinging from the hooks instead of the glittering fish they hoped to see. Cut off and thrown overboard, the cruciform corpses of those albatrosses drifted away, tossing in the wake of the boat.

Holding steady behind longline fishing vessels, some albatrosses also swooped at the baits as the crew was pulling in the line. As the men pulled those wounded birds aboard, they sometimes removed the hook before releasing them. Or, not wanting to deal with large, frightened, menacingly billed oil-spitters, they merely cut them loose and let them flap away, trailing the snood like a wisp. These also probably died later from internal injury or starvation.[10]

Brothers observed that most albatrosses took baits within two hundred yards of the stern and were successful in two of three attempts. Fewer than that got snagged, but he believed more birds succumbed to hook wounds than what observers spotted. Whether an albatross successfully retrieved a bait depended on water conditions and wind speed and direction, which affected sinking rates that kept baited hooks from sinking, as well as on the crew's throwing efficiency, and on whether or not the fishing master was using diversionary tactics. On one voyage, the Japanese master fixed on the upper deck a trolling pole hung with plastic streamers attached at intervals that fluttered out five hundred feet over the sinking line, confusing and scaring the birds and also masking to some extent the sparkling hooked baits until they sank. After the main line had been out the number of hours designated for the "soak," the crew spent the next twelve hours winching it back in, hauling in the submerged tackle, claiming the fish prizes, and stowing each one in ice as it came in. Only big storms altered their grueling seventeen-hour schedule that began before dawn each day.

On seven voyages, Brothers observed and recorded 0.41 birds snagged per thousand hooks deployed, 1.3 birds per vessel each day. But many more deaths likely occurred from internal injuries. Based on his own sample of 108,000 hooks that snagged at least 45 albatrosses during seven voyages and the data supplied to him by Japanese fishers, Brothers published his conclusions in 1991. Longlines set by Japanese tuna fishers in Australia's two-hundred-nautical-mile fishing zone killed at least 44,000 birds annually, he calculated. The actual number could be double that. It was clear that commercial fishing practices were killing large numbers of these giant seabirds in their open-ocean home. Almost half the casualties that Brothers witnessed were Black-browed Albatrosses, followed in descending order by Shy, Light-mantled, Grey-headed, and Wandering Albatrosses. Brothers had seen how larger birds such as Wanderers proved unusually susceptible to longline hooks because they shoved smaller birds aside and were able to gulp down the entire bait and hook, whereas the smaller petrels snapped and picked, breaking off bits of fish and squid as they harried the sinking lines. Brothers also worried about diving seabirds, such as the Grey Petrels, which plunged into the water to retrieve the sinking baits.

By the time Brothers's pathfinding publication appeared, Japanese longliners had almost tripled longline sets inside Australian waters to about twenty-two million hooks annually, compared with a paltry seven hundred thousand hooks set in the zone by domestic tuna fishers. Seven years or so later, additional monitoring aboard Australian tuna fishers showed even higher rates of bycatch, so that even after the Japanese fleet had quit Australian waters, conditions for seabirds had continued to deteriorate. Eventually other researchers linked additional albatross deaths to commercial longlining in other fisheries such as swordfish, halibut, hake, cod, ling, and Patagonian Toothfish.[11]

Brothers's estimate for an industrial longline fishery startled his bird colleagues but not his hosts. Japanese officers claimed that albatross interference with longline baits around Tasmania was not unusually high compared with some areas off New Zealand where more birds came into contact with their vessels. One fishing master indicated that a bigger problem area existed around Heard and Kerguelen Islands and also off South Africa, all islands with albatross populations in decline.

In the meantime, after reading the French speculations about reasons for albatross declines and knowing about Brothers's work and unpublished data, John Croxall wrote a guest editorial for the March 1990 issue of *Antarctic Science*.[12] He blamed commercial fishing in sub-Antarctic waters for overfishing and for dumping garbage from longline vessels, particularly the packing bands used to store bait, which were literally collaring fur seals. Croxall also blamed commercial fishing for albatross deaths. The work of the French scientists, Brothers's work, and his own analysis reinforced Croxall's personal view as to why tuna fishers had reported most of the band recoveries of South Georgia's Wandering Albatrosses after 1975, a practice they stopped after realizing the negative implications. Croxall concluded that longline fishing was a major factor in the losses among Wandering Albatrosses on South Georgia's closely monitored Bird Island. Although some of the Japanese tuna fishers were deploying sea-bird-repelling devices and procedures, Soviet longline fishers off South Georgia were not. Fishing in the bird-filled waters required monitoring and control, declared Croxall, and the body to accomplish this task was the 1980 Convention on the Conservation of Antarctic Marine Living Resources, which held jurisdiction over fishing in the Southern Ocean except for national exclusive economic zones, such as those around the French possessions. However, according to Croxall, the formalized consensus approach to decision making taken by CCAMLR, of which the USSR as well as France, the United Kingdom, and Australia were members, was not working, given "the traditional unwillingness of fishermen to adopt precautionary measures."[13] Members of CCAMLR did not appear, he argued, to think more broadly than their own national interests. Because CCAMLR had failed entirely to study and report on seabird entanglement and incidental mortality, Croxall urged that scientific observers be placed aboard vessels to monitor what was happening to seabirds. In effect, he challenged national colleagues, nongovernmental bird and environmental groups, government agencies involved with management of polar resources, and others, including fishers, to confront the high-seas industry about the need not only for legal, sustainable fishing in the Southern Ocean but also for a reduction in and eventual elimination of seabird mortality. This was a huge task. It would require high levels of political and strategic skills to persuade governmental, nongovernmental, and commercial interests to work together and commit significant resources to solving the problem of birds dying at sea. CCAMLR was the logical starting point because of the overlap of its Southern Ocean jurisdiction with that of the albatross and with Bluefin Tuna and Patagonian Toothfish. At that time, Croxall was a member of the UK delegation to CCAMLR, which meets annually in Hobart, where Brothers and other committed albatross experts were also working. Things were beginning to come together, not a moment too soon. Two years after Brothers's breakthrough article, South African scientists J. Dalziell and M. de Poorter observed dead seabirds being hauled in by crews on two Soviet vessels fishing for Patagonian Toothfish near South Georgia. From that tally, these watchers inferred a catch rate of 0.67 birds per thousand hooks for that deepwater fishery, totaling one thousand albatrosses or more that year, higher than Brothers's estimate from the tuna fleet.[14]

About the same time, in 1991, the Australian National Parks and Wildlife Service, now (September 2010) the Department of Sustainability, Environment, Water, Population and Communities, commissioned a review by zoologist

Rosemary Gales to determine the status of the world's albatross species. Albatrosses at that time were divided into only fourteen species, ten of which were present in the Southern Ocean. With a Ph.D. in zoology from the University of Tasmania, Gales is now section head of Biodiversity Monitoring with the Tasmanian Biodiversity Conservation Branch of the Department of Primary Industries, Parks, Water and Environment in Hobart. She has studied marine bird and mammal species, including seals, penguins, and albatrosses on sub-Antarctic islands. She focuses especially on Australia's own Shy Albatross, which nests on three small islets off the Tasmanian mainland. She and her students have examined its breeding biology and foraging strategy, including how it interacts with fishing vessels. She has plotted band recoveries and color-marked and satellite-tracked individuals to determine where and how far the birds travel from their breeding sites. With Brothers, she became involved in studying albatross mortality from longline fishing and has published extensively on the role of seabirds in the marine environment and on their interactions with fisheries and fishing operations.

Under terms of the 1991 consultancy, Gales reviewed the life histories, fisheries interactions, status, and threatening processes for each species and issued a comprehensive report two years later. By applying IUCN criteria under BirdLife International guidance, Gales determined the status of her favorite Shy Albatross to be officially Vulnerable. Combining this assessment with others being taken for additional species, her report detailed the uncertain and bleak prospects for albatrosses in general. Gales recommended that the Amsterdam Albatross be added to IUCN Appendix I, which included the Short-tailed Albatross as one of the species already believed to be endangered. The report recommended that twelve albatross species not previously listed under the Convention for Migratory Species be included in Appendix II, which lists the migratory animals with unfavorable status that need to be protected through international agreements. She concluded that the populations of only three species had stabilized or were known to be increasing after earlier persecution. Fishing proved to be the common factor in the recent slide, and although only two bird species, the Short-tailed Albatross and the Waved Albatross, had not been reported as being taken in longline fishing at the time, subsequently they have been. Gales emphasized that protecting albatrosses was a "global responsibility."[15] Only through international cooperation and collaboration among the different agencies involved in both fishery operations and management and seabird conservation could the albatross be saved. Gales posed a sobering challenge, but one that we will see has indeed been taken up by a remarkable array of collaborators. The subsequent cooperation and collaboration, on a truly international scale, now also includes

137

members of the fishing industry and the tourism industry, food wholesalers and retailers, championship sailors, royalty and other celebrities, and perhaps most promising of all, the individual consumer. This has culminated in a growing global instrument focused on the conservation of albatrosses and petrels. Gales is the convener of the Status and Trends Working Group of that instrument.

Graham Robertson is another Australian seabird scientist interested in the interactions of seabirds with commercial fishing. Now a senior scientist with the Australian Antarctic Division, he was wintering in Antarctica in the late 1980s to complete his work on Emperor Penguins when he became interested in seabirds, which led him to a doctorate on seabird ecology at the University of Tasmania. By the early 1990s, like Rosemary Gales, he began analyzing seabird deaths as bycatch and joined Gales in her work to expand global understanding of the albatross family and its plight. Robertson continues today to make important scientific and technological contributions to albatross conservation. As early as 1991, before the crisis in albatross populations was fully understood, a group of seven Antarctic tour operators discussed their desire to form a trade association to promote ecologically sensitive travel in that delicate region. Robertson provided interested Antarctic tour operators with the hard information about seabirds and the Antarctic ecosystem they needed to move forward with their plans. He made presentations and distributed CDs, DVDs, and literature to the operators and ultimately was able to sell the importance of albatross conservation to the group. The International Association of Antarctic Tour Operators (IAATO) became a formalized trade association in 1991 and pledged itself to promote seabird and albatross conservation among their guests and devote a portion of their collective profits to albatross conservation. Robertson continues to support IAATO, whose environmental activism and contributions have increased from year to year as the number of tourists in the Southern Ocean has grown geometrically.

The observations and research conducted simultaneously in the Indian Ocean by Jouventin, Weimerskirch, and coworkers on French islands; by Nigel Brothers, Rosemary Gales, and Graham Robertson in Australia; and by John Croxall, Peter Prince, and others on South Georgia began to fit together to make a coherent picture, one that expressed a clear and explicable trend. The albatross had become an endangered victim of advanced fishing procedures and technology, tempted by miles of longlines of baited hooks stretching behind deep-sea fishing vessels. The longline technology that many fishery and other scientists had initially applauded as more efficient in reducing waste of nontarget fish was now proving also to be extremely effective in attracting and entangling an entirely different group of animals habituated to following fishing vessels, especially albatrosses. Albatrosses and petrels were the major seabird bycatch species in the Southern Ocean; Northern Fulmars in the North Atlantic; and additional albatross species, gulls, and fulmars in the North Pacific.

Once these scientists had to some extent quantified seabird bycatch and determined the relationship between longlining and the threatened status of albatrosses, it became clear that the fate of the albatross family was no longer simply a question of bird conservation. The fate of the albatross was inextricably linked not only to the fate of commercial fishing and its bycatch but also to the fate of overfishing and illegal fishing. At the same time that the threatened status of albatrosses was becoming clear, commercial fishers and marine scientists were becoming alarmed at the enormous efficiencies and inefficiencies of their industrial technology and practices. If the albatross fate was linked to the fate of an enterprise of high economic value, and linked in ways that the long-term health of one led indirectly to the long-term health of the other, that tied the fate of the albatross to the market. Finding and quantifying these linkages is considered finding the gold for environmentalists. It was clear that environmental health and economic health were aligned in this instance, as they are in all instances if the view is broad and long enough. The albatross could become a free rider in the economic analysis to save fish. The market principle, based on incentives and reinvestment capacities, would have to be adjusted and recalibrated using a discount rate based on the ecosystem approach, intergenerational equity, and the precautionary principle. No simple task, this, but market incentives have worked wonders in the past. The threat to albatrosses interested and concerned some commercial fishers, but the global threat to fish from overfishing, bycatch waste, and illegal fishing was, by the 1990s, something neither fishery officials nor commercial interests wanted to or could ignore any longer.

At first the two groups operated on parallel paths, the fishery management bodies trying to come to terms with declining and disappearing fish stocks and the environmentalists forming coalitions to deal with the staggering loss in albatrosses as well as sharks and marine turtles. Recognition of the threat to albatrosses catalyzed nongovernmental and governmental conservationists and bird scientists to begin channeling energies into discussions with the governments of fishing nations, commercial fishers, fishery management bodies, and international organizations such as the UN's FAO. It is at this point that our story really takes flight, so to speak. We can track a convergence of forces closing in around the albatross—seabird scientists, environmentalists, government agencies, nongovernmental organizations, fishers and fishery management bodies, and international agreements and conventions. But first it had to be determined if seabird deaths were an unavoidable cost of commercial fishing.

CHAPTER 13

Bycatch

NIGEL BROTHERS CONFIRMED that industrial fishing—longlining for Southern Bluefin Tuna in this instance—was causing the sudden serious declines in albatross populations on Southern Ocean islands. Speculation suggested that as many as three hundred thousand seabirds died each year from being hooked on longline gear and that perhaps one-third were albatrosses. Large numbers reportedly were also dying in trawling, but no trawling counterpart to Nigel Brothers had come forth to quantify the loss. Moreover, albatrosses were also still suffering in their nest colonies from introduced mammal predators and plant species, although governments had been making efforts to restore natural habitat to these lonely islands since Lance Richdale had initiated the first carnivore trapping on Taiaroa Head in 1937. BirdLife International, using data from long-term studies as well as from researchers like Brothers, today classifies 36 percent of albatrosses as Vulnerable, 27 percent as Endangered, and 18 percent as Critically Endangered. No other bird family is currently threatened to this extent.[1]

The term "bycatch" refers to animals caught incidentally and unintentionally in fishing. There is general agreement that commercial fishery bycatch each year tops about thirty million tons worldwide, about one-third of all wild fish landed. Waste among shrimp fishers occurs at the high rate of thirty-five to eighty-five tons per hundred tons caught. Such profligate fishing increases the risk not just to the bycatch species but also to the target fish themselves by killing nontarget species that play essential parts in energy transfers in the food chain, the pathways by which nutrients are cycled. Moreover, when dumped in high quantities, bycatch putrefies and can produce sterile areas on the seabed. The destruction to the seabed from bycatch and from runoff of chemicals from commercial agriculture together have created dead zones in the ocean that we now know are increasing geometrically in number and size each year.

Fish bycatch consists mostly of low-value, undersized, or inappropriate species, but not always. Paradoxically, the quotas that fishery bodies impose also contribute to the waste. A vessel may meet a quota for one high-value target species and then move on to catch a different high-value species, now discarding the first high-value fish as bycatch. Additionally, some vessels high-grade, keeping only the largest specimens of the target fish for market and discarding the rest. Such practices are ecologically disastrous.

The public first became aware of bycatch in the late 1960s and '70s when the media drew attention to the deaths of dolphins caught up in purse seines set in the Eastern Tropical Pacific Tuna fishery. Because fishers realized that tuna often swim beneath schools of dolphin, they set their nets on dolphin. As they hauled in the nets full of tuna, they drowned in the process several hundred thousand Common, Spinner, and Striped Dolphins. The public outcry over these deaths was followed by consumer boycotts of tuna not bearing a "dolphin-friendly tuna"[2] label. This consumer action successfully reshaped the market. Commercial tuna fishers found it necessary to take bycatch seriously and adopted reverse, or "back down," methods when hauling in their nets, a practice that quickly became mandatory. Government intervention through the 1972 US Marine Mammal Protection Act ended the crisis. Instead of resisting the regulations, to maintain their market share, fishers readily adopted the required procedures.

In the 1970s, concern over sea turtle bycatch in shrimp trawls led to a requirement that shrimpers install turtle excluder devices (TEDs) in trawl nets. This helped reduce turtle deaths, but the US National Research Council has continued to list shrimp and other types of trawls as causing "an order of magnitude higher mortality" among Kemp's Ridley and Loggerhead Turtles than any other mortality variable. Whale bycatch from gillnets set for finfish also became a cause for public outcry. International agreements that addressed bycatch, such as those of the International Whaling Convention, the IUCN, and the 1979 Convention on the Conservation of Migratory Species, increasingly pressured fishery operations worldwide. During the 1980s, the large amounts of driftnet bycatch raised more concerns about the costs of bycatch such as unsupportable waste, shifts in ocean trophic structures, and contamination of the seabed.

Because almost all fishery management bodies use single-species population models to determine allowable fish catches, fishing technology and management practices have had little or no regard for the nontarget organisms in the target fish's ecosystem. Fishing methods such as driftnets, dredging, blast fishing, cyanide fishing, and trawling take everything within reach. Driftnets are gill nets extending as long as fifteen to thirty-five miles that float in surface or midwaters to catch squid, tuna, billfish, and salmon. They are so effective they can even yield a profit from seriously depleted stocks. But the free-swinging meshes also entangle "trash" fish, sharks, sea turtles, or marine mammals such as dolphins, seals, and even whales, endangering local populations and depleting stocks. Moreover, before they were outlawed, they snagged an estimated half million surface and diving seabirds per year, including albatrosses. By the mid-1980s, fishing vessels using driftnets in the North Pacific were setting at least 20,000 miles of net daily. By the early 1990s, Japanese operators were hanging out 93,000 miles of driftnet for salmon and 225,000 miles for squid, and protests began pouring in from all over the world. When the UN General Assembly intervened in 1992 to ban driftnet fishing on the high seas, Australia also banned use of driftnets in its waters. Several small island nations welcomed the move because of their concerns about the effects of driftnets on their artisan fisheries. Despite the ban, driftnetting still occurs. As recently as 2003, it was estimated that as many as three hundred thousand whales, dolphins, and porpoises were still dying annually in driftnets.

Dredging severely damages the ocean floor and its living organisms, and dynamiting is equally destructive. In the Philippines, an estimated thirty-eight years are necessary for a 50 percent regeneration of dynamited coral reefs, assuming no intervening explosions. Cyanide fishing was invented in the Philippines in the 1960s to capture reef fish and supply the aquarium trade worldwide. As it stuns the target species, it kills the coral as well as the more vulnerable small fish and invertebrates in the area. Trawling, especially bottom trawling, is similar to dredging and is widely used commercially. It is probably the most destructive fishing method, trapping all nontarget organisms in its path and, even more seriously, destroying the seabed. Midwater purse seining captures fewer incidental species than does trawling. But it was purse seining for tuna that caused the death of hundreds of thousands of dolphins each year until consumer power produced a shift to more selective fishing practices. Fry fishing to supply fish farms is another practice that leads to stock depletion because it targets and harvests juvenile or fingerling fish around certain types of floating objects where fish larvae gather, resulting in what is called recruitment overfishing. Recruitment refers to the availability of juvenile fish to supply the breeding necessary to maintain or restore a population.

Another cause of seabird and other bycatch is lost fishing gear such as pots and traps and the thousands of "ghost nets," or "curtains of death," that drift unattended. They often swirl together into gyres to snag sharks, marine turtles, and seabirds. These untended synthetic nets stretch hundreds of yards and decay very slowly, so they operate for years as navigation hazards for boats and death-traps for fish-eating birds. A so-called rebaiting cycle happens when fish become trapped in the net as bait for larger fish, which become trapped as bait for even larger fish. In the north-central Pacific, US oceanographers and researchers use satellite data to pinpoint areas where these ghost nets come together, still fish-ing. Long-range reconnaissance flights from Honolulu have detected as many as two thousand pieces of debris circulating in whirls of balled-up fishnets. Recent estimates suggest that the so-called Great Pacific Garbage Patch (see Map 1) con-taining these ghost nets and huge pools of plastic, broken and sifted into smaller and smaller soupy fragments, spans an area twice the size of Texas. Some say it is even larger than that.[3]

Initially, people argued that the new longline fishing technology developed in the 1950s was a great boon. It is a fairly selective fishing method, so as commer-cial fishing officially switched away from driftnets, experts applauded the more humane way of harvesting fish such as tuna and swordfish. It allowed fishers to target fish by using specific line lengths and baits, which reduced the amount of fish bycatch. Longlines also compared favorably with trawling in regard to both catch-to-effort ratios and fuel efficiency. Japanese fleet skippers in search of tuna began honing their skills on longlining and worked at it for twenty years or more, until high costs began to push many of them out of the market in favor of other Asian nations. Today, larger pelagic longline vessels operate out of Japan, Taiwan, and the Republic of Korea and land about 83 percent of the tuna catches in the Pacific Ocean, 79 percent in the Atlantic Ocean, and 56 percent in the Indian Ocean. They target five species of tuna. Six additional nations, including the United States, go after tuna, and an additional half dozen countries longline for swordfish in the Atlantic Ocean and Mediterranean Sea.

There is also bottom or demersal longline fishing that targets halibut, hake, ling, toothfish, and other bottom feeders. Fishers harvesting the valuable Pa-tagonian Toothfish now rely on demersal longlining entirely. Fisheries expert Karl-Hermann Kock noted that in the mid-1980s, Soviet vessels switched from trawling to demersal longlining in the Southern Ocean, mainly seeking Pata-gonian Toothfish, and boats registered in Chile and Argentina joined them. A decade later, demersal longline vessels flying flags of a dozen nations were fish-ing there. The demersal longline is a weighted polyfilament line, fitted with per-manent branch lines, that is anchored to the seabed. Vertical lines connected to surface buoys equipped with flags and transponders identify each end of a set.

Underwater deployment may be in the form of straight or U-shaped spreads, and a series of sets may carry as many as thirty thousand hooks. Industrial fleets operate in higher latitudes of both hemispheres where the colder currents preferred by target fish and seabirds well up and circulate over and around submerged plateaus and seamounts. Many fishers use the Spanish system, a heavier safety or mother line from which the thinner longline is hung. If the longline set becomes snagged on the bottom and breaks, the crew haul up the safety line to retrieve the demersal set at its next unbroken linkage. Crews bait the hooks by hand or use an automatic baiting device called an autoliner, shoot the line, allow it to soak for as long as a day, then haul it to remove, bleed, dress, and freeze the fish they bring aboard. They then rebait the hooks, check for wear or damage, and reshoot the line. Big oceangoing vessels stay at sea for weeks longline fishing and often rendezvous with supply boats to offload their catches in what is called transshipment.[4]

As scientists and environmentalists began in the 1990s to recognize longlining's efficiency in killing seabirds as bycatch, praise for its selectivity diminished. People began focusing on finding mitigation techniques and procedures that could be used to reduce and eliminate longlining bycatch. At the same time, details about dead and dying albatrosses began to implicate trawls as well as longlines. The sunken trawl net, pulled behind the boat by cables called the net sonde and warps, is held open with wide wooden or metal boards and dragged either through midwater or just above or even across the ocean floor. It maximizes the net's contact with the bottom and picks up or crushes all the organisms in its way such as sponges, coral, and sea grass. The open end of a trawl net towed by the strongest vessel can span thirteen hundred feet, wide enough to embrace nine Boeing 757 airplanes.[5] Studies by BirdLife International's seabird specialist Ben Sullivan and colleagues in the Falklands and later South Africa demonstrated how cables attached to the trawl were injuring and killing seabirds, notably albatrosses. Videos aboard trawlers showed the cables slicing and crippling birds when they flew in or sat on the water to inspect the net's flashing contents. Albatrosses, with their long wings, appear especially prone to fatal wounds from steel cables yawing with the swells. Some also get trapped in the fish-filled meshes as they lunge to stab into the net. As early as the 1970s, Soviet trawlers taking finfish and krill in the Southern Ocean around South Georgia and Kerguelen were reportedly killing albatrosses. In 2007, observers calculated that twelve thousand albatrosses had recently died in South Africa's trawl-fishing industry. New Zealand's trawl fishery has also taken White-capped Albatrosses. Trawling poses problems for albatrosses and petrels in both hemispheres, including Laysans and Blackbrows in the eastern Pacific Ocean. Factory vessels operating deepwater trawls over the Patagonian Shelf, the Chatham Rise, and the Gulf of Alaska, for example, are likely taking birds in larger numbers than longline vessels. Early on, CCAMLR passed regulations to monitor and manage trawl bycatch in the Southern Ocean. However, compared with our understanding of longlining, the effects of both midwater and deepwater trawling on breeding albatrosses have been difficult to assess, as less data have been systematically gathered about seabirds killed by trawlers.[6]

This has begun to change, however. Procedures are being taken to reduce bird deaths from collision with cables, most notably in fisheries around New Zealand and South Africa. By flying streamers and curtainlike baffles, developed in New Zealand to project behind the vessel where the trawl net comes in, and by keeping fish parts and bycatch on board, crews have been able to reduce the destruction of birds in trawl fishing.[7]

Trawlers attract birds, which strike the cables supporting the net, or are trapped in fish-filled meshes as crewmembers haul it aboard. This is an ongoing research problem. Photo by Robert Grace, © Greenpeace. Courtesy of Greenpeace.

Add the problems of overfishing and illegal fishing to the capacity of long-line and trawl fishing for killing large numbers of seabirds, and the threat to seabirds seems overwhelming. However, there is a hopeful note in the linkage of the fate of the albatross with the fate of commercially desirable fish as well as the marine environment. This fact is what makes saving the albatross more than single-species conservation. Consider the commitments, resources, persuasion, and opportunity costs of efforts to protect the Spotted Owl in the northwest forests of the United States, or the Golden-cheeked Warbler and Black-capped Vireo in the Southwest. And then expand your mental reach to consider the seemingly insurmountable challenges of attempting to protect a truly international and global wanderer such as the albatross. The birds' connection with marine environments that have economic value and importance as the source of food and with World Heritage Sites worth protecting for future generations has provided a solid rationale for an international commitment of time and resources to saving albatrosses. But there are solid rationales for all kinds of beneficial actions on the part of governments. The amazing thing is that once the problem was identified as the connection between albatrosses and commercial fishing, in a relatively short period of time in terms of environmental progress, an international environmental regime took shape to save the albatross. In the ten years from Croxall's editorial in 1990 to the formation of the 2001 Agreement for the Conservation of Albatrosses and Petrels, the efforts of a staggering number of people and organizations at all levels began to emerge as interrelated parts of a complex structure supported by science, increasingly useful data, and significant resources. Earlier we used the metaphor of assembling a jigsaw puzzle to describe this process. It would be more true to say that individuals and groups working independently were shaping the different pieces without at first realizing there were other pieces in the making and a larger picture to assemble. Once they began communicating, they understood that with some recrafting, they could begin to fit the pieces together to make something with the global reach that conservation of a global species required. This is far from a two-dimensional puzzle or a linear story with a single narrative. It's a multidimensional story that is devilishly hard to plot, much more so to narrate, in a way that even begins to capture the hard groundwork from the bottom up and the national, international, and nongovernmental top-down leadership and work that took place, or the complexity, serendipity, and synergy that allowed that work to emerge as an environmental regime for the albatross. Telling that story, trying to give credit where credit is due while most certainly falling short of that goal, is what we attempt in Part IV of this book.

Part Four

SEA CHANGE

CHAPTER 14

Links

THE GREAT ALBATROSS is a powerful icon for saving the oceans and marine life, and there is more at stake than whether or not the albatross survives—not because we can prove that the loss of the albatross is a loss from which the oceans cannot recover, but because our will and energies to save the albatross depend on summoning the same will and energies to save fish, which entails conserving the plenitude and diversity of the entire marine ecosystem. Albatrosses belong to a biological community that extends beyond any nation's jurisdiction. They pass most of their lives at sea—and probably would stay there if they could—approaching land only to nest. Over the past half century or so, commercial fishers have sailed into the biggest swells that encircle the globe at the higher latitudes and hooked or enmeshed thousands upon thousands of the earth's largest winged gliders using the clever technology that extracts valuable fish on an ever more intensive industrial scale. The process has been a disaster.

No fishing method catches only target species, and most fishers prefer not to kill albatrosses. In the final instance, saving albatrosses from death by longlining depends entirely on the goodwill and practices of fishing masters and their crews. But long before saving albatrosses appears on their mental horizons, they have to respond to the need to save fish. Fortunately, pressures and incentives for managing and restoring fish stocks exist, even though some marine scientists and fishers argue that they are too little, too late. To want to save the albatross at this point in its—and our—life history, we must turn to larger, more complex decision-making arenas than that of any individual fisher.

The health and survival of the albatross and fish depend on policies and actions at national, regional, and international levels, in both territorial and high-seas waters. Moreover, their protection and survival ultimately depend on a global culture shift, a sea change if you will, in regard to both catching and consuming fish. But even such complex actions and shifts begin with the work of single individuals, who most certainly work in blind faith that their efforts will lead eventually and over the long, long term to the larger more complex solutions required. We have already met some of those key individuals. The list is never ending, and there are many more than we can introduce here. In the arenas of action we describe, there is room for many more players and many more pieces to be added to the picture.

Political will and power are diffuse and complex. They intersect and spread throughout all human activities at micro levels, take many forms, and practice their effects continually, often without our conscious awareness. A culture that coalesces around a specific theory or agenda can become a juggernaut, as, apparently, our profit- and consumption-driven market has become. Some political theorists suggest that the drive toward totality always trumps more complex solutions, which, because of the saturating effect of a dominant theory or narrative, are always doomed. But we would not have written this book if we believed that the albatross—or the oceans—are doomed. Currently we are watching a system that has been treated as total—the mighty market-based economy—becoming more complicated and open to change as we generate new knowledge about the environment and our interactions with it, especially about those actions that are proving to have been self-defeating to the market system itself. We are at a turning point. As the globalization juggernaut rumbles forward, the mindset propelling and attracting it finds itself in crisis, creating a moment of opening toward new thinking.

Historically, governments have given priority to economic needs and development over environmental protection. In the traditional market-based calculus, harvesting and marketing finfish clearly outrank the problem of saving albatrosses. Self-limiting behavior is not part of the traditional market ethos. But many nations and commercial concerns as well as scientists and environmentalists increasingly recognize the high numbers of fish and albatross deaths for what they are: a waste of genetic diversity and a jeopardy to marine ecosystem integrity and stability, with unforeseen consequences for the market, certainly, but also for life on the planet. There is currently the potential for the kind of value shift that will save the albatross and what it stands for. Such a shift depends on partially unknowable and unpredictable global interactions and collaborations among people, governments, nongovernmental organizations, the media, fishers, and, importantly, individual consumers. These global interactions and collaborations are happening, and they are the source of our hope. However, progress continues to depend on several fundamental components: an independent, adequate, and reliable knowledge base; credible and skillful leadership; equity both in the sense of a new understanding of ownership and in the sense of fairness; and, finally, an increasing coincidence and synergy of environmental interests and political will. All of these, taken together, could lead to the urgently needed culture change.

Not many indigenous groups depend on seabirds, and Southern Ocean albatrosses in particular do not have a large local constituency, except among the Maori in New Zealand. Nevertheless, nations that are host to nesting colonies are assuming leadership in developing both regional and international agreements regarding conservation. Australia, New Zealand, South Africa, France, the United Kingdom, and, increasingly, nations in South America and, less rapidly, Asia, backed by articulate nongovernmental scientists and environmentalists, and armed with data and real-time knowledge regarding dwindling bird populations within national jurisdictions, are setting an example and demonstrating effective leadership among the regional and international organizations to which they belong. They offer positive role models for similar range states to draw upon in reducing bird losses, both at sea and on land. Successful efforts by the United States and other countries to shift the International Whaling Convention from a benighted sustainable harvest agreement into the protection body for whales it is today demonstrate the power of leadership from within. The terrible irony of that, however, is that the shift toward protection occurred after commercial whaling had virtually extirpated the stocks of blue, right, and other baleen whales.

151

In light of the inequities between north and south, it is interesting that the so-called group of seventy-seven, the first states to ratify the 1982 UN Law of the Sea Convention, which went into force in 1994, consisted mostly of developing nations, while many Northern Hemisphere nations stood back.[1] So far the United States has signed but not ratified the Law of the Sea. As we saw, it was coastal states in South America as early as the late 1940s that first asserted authority to manage living and nonliving resources off their shores, including migratory and boundary-straddling fish stocks on a regionally flexible basis through bilateral agreements. The intrusion of distant nations' fishing fleets into their rich waters threatened these states' artisan fishers and their villages and towns. Because Law of the Sea negotiators recognized the effectiveness of nations' management of their own coastal waters, the eventual agreement supported and broadened national authority over territorial waters, national fleets, and nationally flagged vessels on the high seas. As a result, nations have increasingly moved to draw up and enforce regulations to manage resources within their own waters. Some commercial fishers and fishery management bodies read this as showing that those with a substantial, material interest in outcomes are best at practicing sound management. Many are taking such an approach under consideration for fishery management.

A strong multinational constituency has developed for the protection of the Southern Ocean as well as Antarctica. Nations and nongovernmental organizations in areas outside Antarctica and its encircling seas are also becoming interested, self-identified stakeholders in what happens there. On the grounds of heritage, intergenerational equity, and a more knowledgeable understanding of the interconnectedness of marine ecosystems worldwide, these players are now collaboratively appealing to a broader world community for shared responsibility and accountability.

The network of bilateral, multilateral, and international accords and treaties that has grown over the past forty years to protect oceans and their resources on a sustainable basis is encouraging. The 1980 Convention on the Conservation of Antarctic Marine Living Resources exercises a broad range of international cooperation and oversight. It also forthrightly supports albatross survival. In agreements like the 1979 Convention on the Conservation of Migratory Species of Wild Animals, the 1995 Fish Stocks Agreement, and the 1991 Code of Conduct for Responsible Fisheries, stakeholders with disparate and seemingly conflicting interests have come together to do the hard work of defining shared principles and drafting soft and hard laws that may eventually work effectively to stop human destruction of life at sea. The 2001 Agreement for the Conservation of Albatrosses and Petrels (ACAP) most probably would not have occurred without the support and leadership of the Australian and New Zealand governments and experts from similar nations, from their polar and marine agencies, and from nongovernmental organizations such as BirdLife International. The fact that both Peru and Ecuador have signed and ratified the albatross and petrel agreement demonstrates the ability of stakeholders to more fully grasp linked issues, such as seabirds and commercial fishing, as they begin to monitor regional and global ecosystems in order to protect and further their own economic interests. By calling for its convention to be run on a basis that recognizes economic differences among nations, ACAP invites and welcomes global participation by nations at all levels of development. Whether coalitions of states and nongovernmental organizations can effectively protect equity and requisite diversity in regard to environmental matters, such as the long-term viability of albatross populations, remains to be demonstrated. The fact that a number of sub-Antarctic albatross nesting islands are now official World Heritage Sites is certainly a step forward for the birds.

Regional and international rules and regulations most often fall under what experts term "soft law." That is, they express intentions and guidelines rather than legally binding laws expressed in treaties or conventions. Soft law is a category that the parties understand and generally agree to within a specific practice. Like an intention expressed through a handshake, parties risk loss of credibility and political face should they fail to live up to their publicly stated intention in regard to policies and behaviors that are neither imposed nor legally binding in the restrictive form of a treaty. Some scholars in the growing and complex field of environmental law welcome soft law as a necessary part of a hardening process whereby intentions may evolve over time into more explicit, legally binding and enforceable provisions of a treaty. Others, however, regard soft laws as delaying or obfuscating tactics that mask the ongoing existence of threats to species and ecosystems. Some also describe the process, not entirely incorrectly, as a willful practice of cynicism. Nations formalize positive and concrete terms on paper despite knowing that the domestic political will does not exist to enforce such agreements and despite recognizing the near impossibility of monitoring what happens on the high seas. In effect, soft-law agreements mislead us to think things are going in the right direction while masking a continuation and even intensification of the underlying threats to species and ecosystems. For example, measures taken by CCAMLR to lower the incidental take of seabirds are the most effective and comprehensive to date and are an essential step toward reducing the transboundary costs borne by albatrosses and their nesting nations. But despite the existence of these well-designed regulations, some nations that helped draw them up and approved them do not comply with them, while others are completely indifferent to issues like bird bycatch. Inducing, persuading, shaming, or forcing participants into compliance is the challenge for a fishery management body like CCAMLR, with a consensual form of decision making that reflects, as one person puts it, "the law of the least ambitious programme."[2]

Whether a nation-based or convenience-flagged fishing industry alters or adjusts its practices to conform to regulations established by regional fisheries or multi- and international conventions requires constant discussion, persuasion, motivation, and enforcement. But these top-down efforts by themselves are insufficient. They have to be met and matched, possibly even outreached, by bottom-up responses and efforts, which takes us full circle, back to the minds and hearts of individual fishers. Encouragingly, some fishers are increasingly prepared to accept measures that lower catches in order to protect the shared resource, usually a fishery, but increasingly also a habitat or ecosystem. For example, a large southern swath of the Indian Ocean is now a whale preserve, and the valuable Patagonian Toothfish fishery around South Georgia closes when albatrosses are nesting to give them a chance to breed successfully and build up numbers. As a result, by 2000, more than fourteen million hooks in that South Georgian fishery killed fewer than fifty birds, or .0035 per one thousand hooks, well below the target set by Australia of 0.05 per thousand.[3]

In some fisheries, operators have been able to agree on a system of ownership or rental rights that gives the owner or renter a strong incentive to protect the long-term viability of the stock. This is not the same thing as assigning quotas based on the calculations of a management model. Quotas provide no incentives to resist certain quite wasteful practices. Fishers fulfill one species quota and then switch to fill a quota for another species in the same area, in the process turning the first target fish into bycatch. Quotas also encourage the wasteful practice of high-grading. On the other hand, fishers who purchase rights to fisheries do not have the same incentive to maximize their take, preferring to use their own best judgment about how to protect the resource for long-term profitability. The downside of fishery rights, property, or rental is that a secondary market develops for trading these rights, so that they eventually accumulate in the hands of fewer and fewer commercial fishers. This can have unforeseen ripple effects so that Hardin's "tragedy of the commons" scenario is replaced by what is in effect no longer a commons, but private property. Nevertheless, properly designed, a system of resource rental rights holds promise for the sustainable use of shared resources.

Identifying what linked issues offer the broadest coverage of both economic and environmental concerns allows collaboration among nations, nongovernmental organizations, fishery bodies, fishers, and multinational organizations for management that does not stop at one species or animal group. Using the concept of linked issues to tie an environmental concern such as albatross survival to the goals of regional planning and economic development can lack persuasive power. But focusing discussions on the costs of depleted fish stocks and illegal fishing may provide an incentive for stakeholders to collaborate to identify and penalize illegal fishers, who are also key agents in destroying seabirds. Discussions about reducing illegal fishing acknowledge the need for cooperation between commercial and artisan fishers to identify and target the illegal practitioners. They also emphasize the need to train and subsidize observers on vessels in national waters and on the high seas so that accurate catch and bycatch data are consistently reported from these sources. Developing the means of funding new technologies, such as satellite-based surveys, and making more reliable fish stock and bird bycatch assessments to assist states in their vessel licensing process are both crucial. All of these activities work in the albatross's favor.

By 1991, several major pieces of the albatross puzzle had begun to emerge and fit together to make a more comprehensive plan, or map, for saving the albatross. Scientists had identified incidental bycatch mortality in commercial fishing as the primary cause of the precipitous drop in albatross populations after the early 1960s. They had quantified the problem and, along with engineers and fishers, were working on technical, procedural, educational, and training measures to stop it. The fishery body with ecosystem management responsibility for the entire Southern Ocean, after a slow start, was agreeing on and implementing important regulations to govern commercial fishing in its area, including seabird bycatch.

By 1993, CCAMLR, with urging from John Croxall, had formed a Working Group on the Incidental Mortality of Albatrosses from Longline Fishing (WG-IMALF; see Appendix). CCAMLR was still hampered by its consensus rule for decision making, and no one had yet understood the contribution of trawling to seabird deaths. BirdLife International, formed in 1993, was building a global partnership and developing a global program for saving seabirds. What was not yet in place was an overreaching structure for placing all the pieces into a larger and more coherent program for addressing what Croxall regards as a "high-seas mess." People had realized that saving the albatross required nothing less than dealing with that mess. Fortunately, the albatross would benefit from all the linked issues having to do with global nutrition, ecology, and economics. But to move forward in solving the albatross part of the puzzle, it was necessary to identify and develop cost-effective devices for reducing seabird deaths as fishing bycatch.

155

CHAPTER 15

Engineering

NIGEL BROTHERS'S ONBOARD observations and discussions with Japanese and other nations' fishing crews and masters led him to begin testing and comparing various techniques and procedures to reduce and eliminate bird deaths in longline fishing. He also continued working to quantify the cost—in time, effort, and lost fish and fuel—of hooking seabirds. He had to find equipment that was cost-effective for fishers to use. Moreover, without efficiency gains, bycatch mitigation would hold little or no interest for commercial fishers. He eventually was able to argue to fishing captains and commercial interests that catching birds reduced profits. And he generally found that, while not always agreeing with him, they were at least willing to give him a hearing.

More than a dozen measures have proven to be effective, both singly and in various combinations (Table 4). During his onboard observations, Brothers had seen some Japanese skippers using bird-scaring lines, called *tori*, but in his 1991 article, he argued that "the simplest solution is to confine line setting to night time."[1] Demersal fishers seemed to agree. In response to complaints about bird losses in New Zealand and Australian waters, Japanese pelagic longliners had also begun setting lines at night. They believed it reduced bird deaths anywhere from 60 to 95 percent, depending on the brightness of the moon and how strong summer twilight held in polar latitudes. However, little data showed increases in the numbers of tuna caught, and some believed the birds would become accustomed to night setting and swirl around the vessels as they did in daylight. Brothers discovered that some fishing masters were towing scare buoys, setting out artificial lures, or setting off explosives to scare the birds away. He also noticed that the practice of adding weight to the branch lines, first adopted in the Indian Ocean, removed baits more rapidly from the birds' view.

Although Japanese vessels rarely used bird-scaring lines prior to the late 1980s, they did employ them in the Australian Fishing Zone under terms of their 1979 agreement (Map 4). By 1995, all seventeen observer-boarded Japanese vessels in the Australian Fishing Zone flew bird-scaring lines, and two years later, the Japanese Fisheries Agency ordered its longline fleet to use them in all oceans. Australian calculations estimated that the billowing streamers reduced bird losses by 30 to 70 percent, although larger and bolder birds still plunged right through them. The lines were also less effective in pelagic fishing because the longer branch lines with lighter hooks sank more slowly, remaining visible for three hundred feet or so behind the vessel.[2] Australia and New Zealand began requiring the deployment of streamers in their longline fisheries, as did the

TABLE 4. Summary of Seabird Bycatch Mitigation Measures

MEASURES	FISHERY TYPE DEMERSAL/PELAGIC		DEVELOPMENT STATUS	COST	SEABIRD MORTALITY REDUCTION
Weighting longline gear	Good	Moderate	Partly	High startup	Excellent
Thawing bait	Poor	Good	Some	Low	Moderate
Line-setting machine	Moderate	Moderate	Developed	Moderate	Moderate
Underwater line setting	Moderate	Moderate	Underway	High startup	Complete
Streamers	Good	Moderate	Developed	Low	High, variable
Bait-casting machine	None	Poor	Developed	High startup	Moderate
Brickle curtain for hauling bay*	Good	Good	Developed	Low	High
Artificial baits	Poor	Poor	Concept	High startup	Good potential
Hook modification	Poor	Good	Concept	Moderate	Moderate potential

Source: Brothers, Cooper, and Lokkeborg 1999.
*A bird-scaring curtain of streamers placed on both sides of the trawl hauling ramp.

Commission for the Conservation of Southern Bluefin Tuna in its area. During the 1990s, however, Brothers and colleagues noted that most Australian fishers were not flying bird-scaring devices even though mandated. Deploying them prior to line setting meant extra work for the crew, and the streamers, depending on their positioning and the wind direction, could easily tangle around the longline. But the use of bird-scaring lines eventually became mandatory among pelagic and demersal fishers in the Southern Ocean, as did night setting south of 30°S for all domestic tuna fishers. They found that when bait-throwing machines were used, paired streamer lines worked better.[3]

Bait-casting machines increase sinking rates by throwing hooks away from wake turbulence and, used with streamers, can reduce casualties, some say, by 75 percent for pelagic longliners. Because thawed fish or squid sink more rapidly than frozen ones, baiting the hooks with thawed baits also reduced the time birds could see them, and using thawed bait is now customary in fishing. Casting machines using thawed fish baits with their swim bladders punctured to reduce buoyancy also lowered seabird casualties by as much as 80 percent in pelagic longlining. However, the use of autoliners, machines that bait hooks mechanically, resulted in more albatrosses being snagged. The brief interruption in line release caused tension to build that doubled the line's sink rate.[4]

Breaking birds' associations with fishing vessels by stowing processed fish and keeping discharges to a minimum effectively lowers casualties. If there is no room to store fish bycatch, crews are urged to throw it overboard, minus fish hooks, when lines are not set and always away from the side of the set. Environmentalists discourage pelagic seabird tour boats from the practice of chumming to pull seabirds in. The aim of all these advances is to wean seabirds entirely from association with fishing boats.

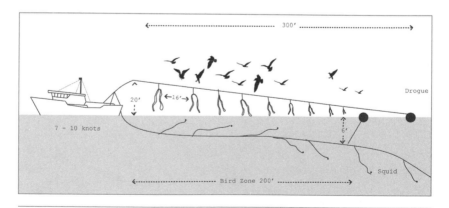

FIGURE 2. A longline vessel flying streamers to baffle seabirds. Drawing by Anwar Slitnine, adapted from Melvin 2000 and Australia, Department of Agriculture, Fisheries and Forestry 2003.

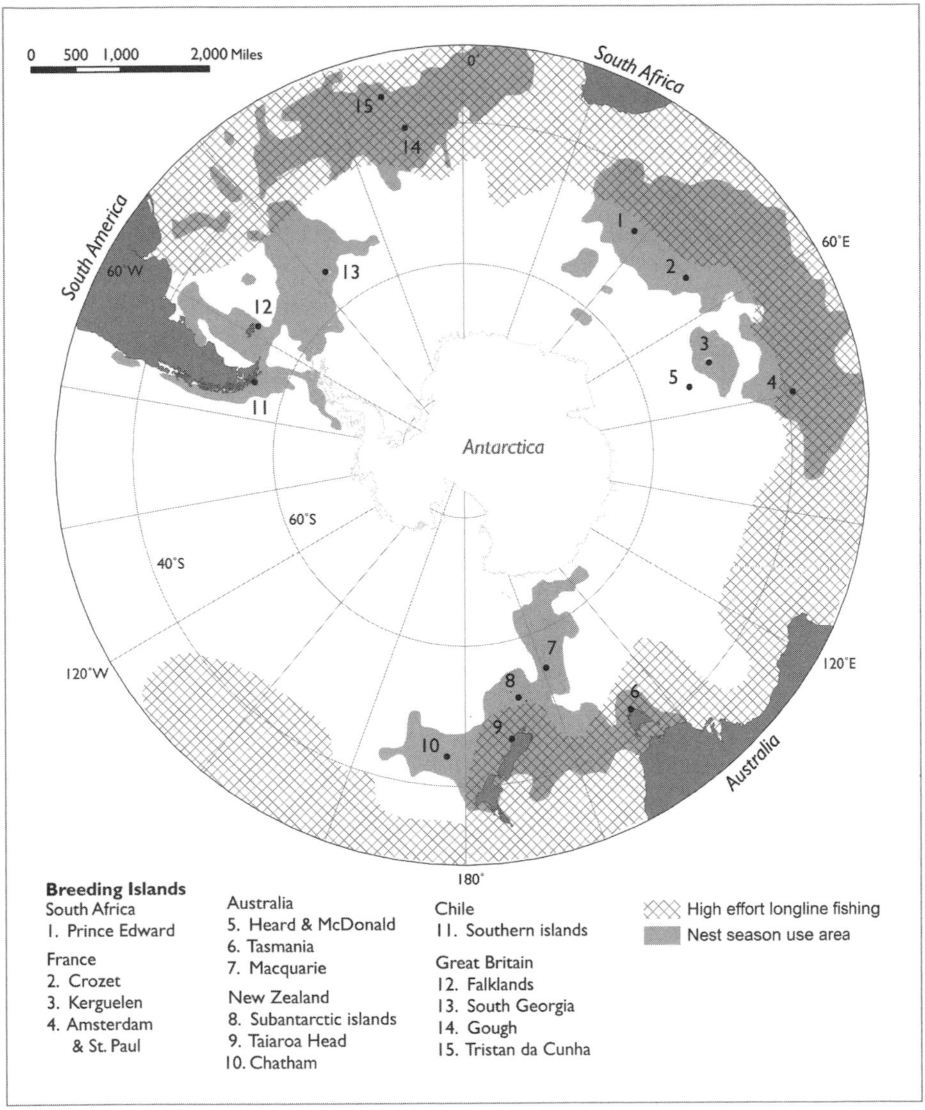
0 500 1,000 2,000 Miles
0°
South Africa
South America
60°W
60°E
Antarctica
60°S
40°S
120°W
120°E
180°
Australia
1
2
3
4
5
6
7
8
9
10
11
12
13
14
15
Breeding Islands
South Africa
1. Prince Edward
France
2. Crozet
3. Kerguelen
4. Amsterdam & St. Paul
Australia
5. Heard & McDonald
6. Tasmania
7. Macquarie
New Zealand
8. Subantarctic islands
9. Taiaroa Head
10. Chatham
Chile
11. Southern islands
Great Britain
12. Falklands
13. South Georgia
14. Gough
15. Tristan da Cunha
High effort longline fishing
Nest season use area

MAP 4. Longline Fishing and Albatross Feeding Areas. Areas of high-intensity longline fishing and bird use areas derived from BirdLife International 2004b. Longlining occurs at low intensities farther south.

Newer technology for setting lines underwater holds promise. The Mustad funnel, an attachment that shoots the line from the stern and frees hooks five feet below the surface, is commercially available for single-line demersal fisheries. Many petrels can easily dive to that depth, however. The manufacturer has lowered the depth for the device, arguing that money saved on lost time and baits offsets the cost of the funnel. Another device uses an underwater chute to set the line. Bycatch rates decline significantly when these devices are used, depending on water conditions and the amount of propeller turbulence, which can force baited hooks back up to the surface.[5] Trials with a deeper setting chute, designed by New Zealand fisher Dave Kellian and developed for pelagic use by Australian and New Zealand fisheries, delivered hooks at twenty feet and even deeper, "like a sewing machine but with a ten-meter-long needle," Kellian says.[6]

After leaving his job as a state biologist in Tasmania, Brothers assisted in developing and testing the underwater chute. Members of the Hawaiian tuna longline association liked the results, but another ten-vessel trial for tuna longlining during 2001–2002 and additional testing off Australia's east coast found that it didn't work for all seabirds. Flesh-footed Shearwaters, expert divers, continued to grab the baits and hook themselves, even at a release depth of sixteen or so feet.[7] So far, however, even underwater setting devices do not reduce bycatch below the elusive target set by Australia of 0.05 birds per one thousand hooks. The rough seas in the Southern Ocean and malfunctions can make chutes erratic. They are also expensive, and experts are reluctant to recommend them for general use. Nevertheless, in 2009, an Underwater Baited Hook team of Aussies and Kiwis, using Killian's prototype, hydroengineering, and biological skill, received the annual Smart Gear Award for consistently setting hooks below the curtain of propeller turbulence and thus away from seabirds. One newer development in fishing boat design is for the line-shooting apparatus to be built into the vessel's keel so that the line never appears above water as it is being set.[8]

Australian and New Zealand fishers continue to develop and test new bird-friendly fishing equipment. Kiwi snapper fisherman Alex Aitken trailed fish oil under the streamer lines as he fished because he had noticed that seabirds swerved from the oil slick, halting seabird losses in his gear. Australian tuna fisher Peter Robinson recognized the same results. For this simple, cheap, and effective deterrent, in 2004 Aitken and Robinson shared BirdLife International's annual Competition of Ideas Award of 18,000 Euros, an award sponsored by BirdLife's Spanish partner.[9]

Japanese authorities have approved a simpler method, that of camouflaging baits by dying them blue (see Table 4). Opinions differ about the effectiveness of this procedure. Some believe the birds have the ability over time to spot dyed baits. Using them is, however, more effective than other methods Japanese vessels have tested, such as firing water cannons or interfering with birds' magnetic fields. Over the past decade, Australian Graham Robertson and American Ed Melvin have been studying how different weights attached to lines affect their sink rates. In 2007, the Pacific Seabird Group awarded Melvin a special achievement award for his work in making fishing gear less dangerous to foraging birds, including albatrosses. A prestigious three-year Pew Foundation Fellowship in 2004 enabled Robertson to systematically test differently weighted line sets off southern Chile, a place where Black-browed Albatrosses were drowning in demersal longlining for toothfish and hake. He tested a Norwegian line with weights integrated into the line that disappeared immediately as it hit the water.[10] From an albatross-supporter's viewpoint, lines can't sink fast enough, and anything that hides baits from inquisitive birds is worth investigating, but there is a trade-off in weighting lines. Weighted lines enter the water and go to work faster, but lighter and longer lines with equally light but tough hooks can carry larger numbers of baited hooks. Although the debate goes on about how much weight is needed for optimal fishing, and where it should be on the main line, integrated weighted line is now the industry standard. A combination of devices works best, such as night setting or *tori* lines with extra weights on the line to make it sink speedily once released.[11]

Fishing around seabird nest islands has factored importantly in albatross declines. Commercial fishers have closed or delayed opening seasons when they found that fish stocks had been depleted. But only recently have fishery management bodies been able to more systematically close fisheries seasonally, based on better science, to allow stock recovery and keep seabirds free of entanglement. In this practice, CCAMLR once again has been a leader. A number of range states have identified areas of high seabird mortality and called for seasonal closures in their National Plans of Action to reduce bird losses. For example, the United Kingdom prohibits longlining when seabirds are nesting on South Georgia. However, it has been observed that seasonal closures merely shift the fishing effort to adjacent seas, with no net gain in conservation.

Brothers and colleagues realized that lowering seabird bycatch also required fishing crews that knew about the problem and took it seriously. Fishers see hundreds of seabirds around their vessels, and it is understandable that they might not relate drowning one albatross or two on a set to a threatening decline in overall numbers. Brothers realized quickly that the education and training of vessel personnel was a critical mitigation procedure in itself. Fishers' pay depends on the quantity of fish they catch and on market conditions when they get back to port, so time is critical. Generally, they consider dealing with a snagged or tangled albatross a waste of time. Moreover, they resent the publicity and condemnation from nonfishery folk who criticize methods long honed at sea for catching valuable fish quickly and efficiently.

Brothers's work in the mid-1990s led him to conclude that the use of one or more streamer lines in combination with crew awareness and training lowered albatross mortality by as much as 95 percent in pelagic operations and 100 percent in demersal operations. Brothers also calculated that the Japanese fleet's use of streamer lines saved about $5 million annually in time, effort, and fish lost to hooked albatrosses. Accordingly, he began a training program for crews and fishing masters through pictures, pamphlets, and videos. He worked from the premise that low-paid, hard-working crews do not want to snag albatrosses, although as with other bycatch species, such as sharks or sea turtles, some seabirds may prove useful as food or as extra baits. He urged crewmembers to be more bird conscious, emphasizing savings in bait, equipment, time, and fuel. Importantly, he solicited and welcomed information and suggestions from fishers about how to improve the situation. Brothers authored and distributed pamphlets describing low-cost, easily implemented techniques to avoid snagging birds. His 1995 publication *Catching Fish, Not Birds* has appeared in more than four languages, and another, *Longline Fishing: Dollars and Sense,* was published in English and Spanish. Written originally for the Japanese longline fleet, *Catching Fish, Not Birds* received the backing of the Japan Tuna Federation International, and the president of the Tuna Boat Owners Association of Australia recommended it as a "big practical advance for the fishing industry."[12] In 1995, Brothers participated in demersal longlining for Patagonian Toothfish off the Falkland Islands to better understand the specifics of the seabird bycatch problem in that fishery and provide guidance to fishing masters. Brothers's documented observations and recommendations have influenced fishery management decisions throughout the Southern Ocean, and he has trained a large number of crews to reduce albatross deaths. However, the constantly high rate of turnover among crew members works to dilute onboard awareness and expertise.[13]

After his 1991 publication, Brothers teamed up with Rosemary Gales to broaden the base of stakeholders involved in preventing seabird deaths from longlining. They briefed colleagues, officials, and politicians about the threat that modern fishing methods posed to seabird species and then took the problem worldwide in the turning-point year of 1995, when they held the first World Albatross Conference in Hobart.

CHAPTER 16

Turning Point

IN RETROSPECT, WE COULD CONSIDER 1995, and the decade of the 1990s, a turning point, and not just for albatrosses. We don't know if the correction was sufficient, or in time, but beginning in 1995, key individual, institutional, commercial, and NGO players stepped forward to take action, with remarkable results. They formalized a global knowledge base for the albatross and for bycatch mitigation, created the necessary structures of international law to deal with the messy situation on the high seas, established an environmental certification system for fisheries and retailers to empower the consumer, and globalized training and bycatch mitigation efforts throughout the world's fisheries. During the next decade the albatross became famous as celebrities such as Britain's Prince Charles, the America's Cup yachting teams, the Volvo racers, John and Marie Christine Ridgway, Dame Ellen MacArthur, Sir David Attenborough, Taiaroa Head's "Grandma," Queen Noor, Jerry Hall, Olivia Newton John, Brian May, and Frankie Dettori got on board to save the albatross.

Multiple events occurred simultaneously across the fields of science and engineering, education and training, international conventions and laws, industry innovation, and nongovernmental policy and action. One was Brothers's and Gales's first World Albatross Conference, the result of imaginative and global leadership on the part of these two Tasmanian scientists. Others were the outcomes of years of negotiations and consultations undertaken by FAO's Committee on Fisheries. Another was an idea that occurred simultaneously in 1995 to two leaders in business and conservation. And another was a program launched by the world's largest and most effective bird conservation group, the global partnership called BirdLife International.

These events, taking place across such a broad field of interests and commitments, clearly did not just coincidentally and spontaneously spring up in a cluster around 1995. It's not possible to tell, or even know, the whole story, but it's clear that those events made saving the albatross a matter of global concern and action available to everyone. It was the commitment of untold numbers of individual people working for what they believed, in private and public venues, that led in unpredictable and unknowable ways to the emergence of an international environmental regime for the conservation of albatrosses and fish in or around 1995.

We've seen how key scientists, working separately on widely spaced Southern Ocean islands, simultaneously identified the steep declines in Southern Ocean albatrosses. We've also watched Nigel Brothers and his colleagues doing the hard onboard work of observing and quantifying the problem and then of working to develop the equipment and procedures to reduce and eliminate it. Without hard data, it would have been impossible to harness the international energies that came into play to save the albatross, and Brothers provided the initial scientific analysis of the threat to albatrosses and its causes. John Croxall, as a member of the UK delegation to CCAMLR and one of its leading scientific advisers, effectively argued that CCAMLR held the primary and most legitimate authority for working to resolve the problem of seabird bycatch. By the time of his 1990 guest editorial, CCAMLR had only begun to acknowledge that seabird bycatch mortality was a problem. There was then no hard data on numbers and causes of the deaths, and CCAMLR had taken no action to investigate the reasons for the large numbers of deaths—thus the ire with which Croxall criticized CCAMLR in his article. At the same time, Rosemary Gales was doing her part, commissioned by the Australian government to conduct a thorough scientific assessment of the global status of albatross populations. Her report provided another key component to an effective analysis of the situation.

Once Brothers and Gales had published their research results, in 1991 and 1993 respectively, the emerging picture seemed pretty clear, even with the limited reach of their studies. The Australian Antarctic Division and the Tasmanian Parks and Wildlife Service then recommended that Brothers and Gales coinvestigate major problems concerning albatrosses and fisheries. So Gales and Brothers took a step that allowed the first truly global analysis of the problem and one that laid the groundwork for developing global strategies to resolve it. They realized they needed a global knowledge and stakeholder base in order to move forward in preventing seabird deaths from longline fishing. To this end, they organized and convened the Inaugural World Albatross Conference, held in Hobart in September 1995, supported by the Tasmanian Parks and Wildlife Department and the Australian Antarctic Division. Approximately 120 seabird and fisheries experts attended, from Alaska, New Zealand, Australia, the United Kingdom, France, the United States, Ecuador, South Africa, Argentina, Brazil, Uruguay, Canada, Belgium, and Japan. Participants included Tickell, Croxall, Jouventin, Weimerskirch, Chris Robertson, and Hiroshi Hasegawa. Presentations summarized research findings worldwide, including a workshop report by Gales, Graham Robertson, and Karen Alexander on incidental mortality of albatrosses from longline fishing. If there was any doubt about the seriousness of the threat to the albatross family, the seventy-three presentations at the conference made it dramatically clear. Scientists were prepared to implicate longline fishing in population decreases measured over twenty years for some species, and declared that "it is likely that all albatross species are affected by this fishing practice."[1] The conference drew together widely separated seabird scientists, fishery management officials, and interested fishers into an energetic collaborative network focused on saving the albatross. Because of its own clearer understanding of the problem, the Australian government further dramatized the threat in another world arena by nominating eleven additional albatross species for the Appendices of the Convention on Migratory Species a little over a year after that first albatross conference. Gales and Brothers believed that such a nomination would not have been possible without the bigger picture the scientists at the Hobart gathering had begun to put together. In 1998, Robertson and Gales followed the conference with their edited publication of *Albatross: Biology and Conservation*, a collection of the research results and findings of scientists around the world. Consisting of thirteen chapters drawn from conference presentations and ten additional chapters solicited to round out the survey, the book covers population studies and trends for all albatross species, foraging interactions with fisheries, mortality due to fishing and other factors, and an assessment of current albatross research and conservation.

A second Albatross Conference, held in Honolulu, Hawaii, in 2000, expanded discussions to petrels and was followed by a third global meeting in Montevideo, Uruguay, in 2004. Brothers and Gales also began discussions with members of the fishing industry, beginning with a paper they delivered to a working group of the Commission for the Conservation of Southern Bluefin Tuna, a major fishery management body pertinent to both tuna and albatrosses. It was a small step but a symbolic one, they thought, that a fishery organization tightly focused on stocks of a single species would pay attention to the unintentional side effects of its tuna longlining operations. Gales and Brothers also collaborated with other governmental conservationists, like the Australian Nature Conservation Agency's Canberra-based Barry Baker. Along with Baker, they promoted efforts within the national government as well as among bird specialists worldwide that fostered the seemingly unlikely development of an international environmental regime for the albatross that is in place today, still wobbly and missing key pieces but structurally holding up.

These scientists have been pioneers in using every available means to expand the reach of biological research and its practical applications in conservation. Rosemary Gales continues studying seabird biology, especially in Australia. She coordinates global reviews of the status of albatrosses and, as a specialist on seabird mortality in fisheries, works like Brothers to monitor, quantify, and analyze the nature and extent of seabird mortality associated with longline fishing. Through their commitment and that of their colleagues around the world, Brothers, Croxall, Gales, Robertson, and Baker were able effectively and in a relatively short time to draw the attention and involvement in seabird bycatch mitigation of state, national, and international bodies charged with the oversight of marine resources. They described the nature and extent of the problem to the stakeholders and laid the groundwork for finding solutions. Developing solutions depended on three kinds of work: applied science and engineering to quantify the problem and to develop mitigation techniques; education and training of the people who did the work, the fishers and their fishing masters; and global strategic thinking and action. Everyone realized that nothing less than an international environmental regime would be necessary to save the albatross, but that must have seemed an unlikely outcome in the beginning. The only existing multinational regime to protect a single species was the one established for the polar bear, and that initially must also have seemed unlikely. The albatross, however, was the beneficiary of an environmental regime that was developing at the same time to save wild fish.

As a global organization, the United Nations has pressed coastal nations to develop policies and implement plans to ensure sustainable fishing. The framing convention for this work is the 1982 UN Law of the Sea. The UN held the first of three conventions on the Law of the Sea in 1956, when there were growing concerns about the rapid expansion of commercial fishing. Nations had strong interests in defining generously the territorial waters over which they had jurisdiction, but little came of the first two conventions. The United States and the Soviet Union dominated these gatherings, and national claims over territorial waters conflicted without recourse to a higher standard or law. In 1973, the UN convened the Convention on the Law of the Sea III (UNCLOS III), with over

160 participating nations. In twelve sessions over nine years of negotiations and twelve more years of signing, the required number of signatory nations entered the Law of the Sea into force in 1994. Fifteen years later, it had 160 ratifications. The Law of the Sea codifies national territorial jurisdiction as extending 12 nautical miles from the nation's baseline, usually the low-water line, and an additional twelve nautical miles as a contiguous zone in which a state continues to enforce laws regarding certain types of activities. It also establishes a method for defining the territorial waters of archipelagoes. The Law of the Sea creates a hard-law legal basis for a larger national exclusive economic zone of 200 nautical miles from the shoreline—or to the median between two countries if national waters overlap—with a monopoly on minerals to 350 nautical miles for coastal nations. It further designates the outer edge of the continental shelf, or 200 nautical miles from the coastal state's shoreline, as the natural extension of the nation's land territory. Additionally, the Law of the Sea established a formal procedure for settling disputes over waters and their marine resources.

The jurisdictional and economic zones UNCLOS defined are essential to our story. Collectively, exclusive marine economic zones encompass 36 percent of the sea surface on the globe, affect 159 states, and provide 90 percent of commercial fish and a similar percentage of offshore oil reserves worldwide.[2] As we've noted, the nation is the only enforcer of international law, and the coastal nations are the best protectors of marine resources in which they have an economic and philosophical stake. As a result, the UNCLOS acknowledgment of a two-hundred-mile national marine jurisdiction produced greater potential for sound environmental management than had previously been possible. Although the United States has not ratified Part XI of UNCLOS, which establishes an international seabed authority outside national jurisdiction, like most other signatories, it considers most of the remaining provisions as binding.

By definition, all waters beyond these territorial limits are the high seas. UNCLOS states that the high seas are open to fishing by all states and that all states have the right to flag ships, but there must be a "genuine link" between states and the ships they flag, which, as we have pointed out, is often not the case. Article 117 imposes a general duty on all signatory states to "take such measures for their respective nationals as may be necessary for the conservation of the living resources of the high seas, and states must cooperate, both regionally and globally, to develop international rules and standards to protect the marine environment."[3] As the 2003 OECD round-table paper notes, however, although UNCLOS established the flag state as the enforcement mechanism for the obligations of its flagged ships, "it is not states but fishing boats that go fishing."[4]

Moreover, the UNCLOS protection of flag-state rights is not accompanied by any legal definition of their responsibilities, nor has there been a willingness to define what a "genuine link" between the vessel and its flagging state must be.

Unfortunately, even the states that are participants in and signatories to international agreements as well as regional organizations like fishery management bodies are not necessarily rigorous in ensuring that their flagged ships comply with those agreements and laws. Some quite blatantly ignore them. Moreover, many flagging states are not signatories to these agreements, which means they are not bound by the measures in the first place. The reality is that many states either cannot or will not take enforcement action against fishing boats flying their flags even when their activities are clearly damaging to the marine environment. Bluntly put, they are happy to claim the rights and benefits of sovereign states to confer flag status without accepting the correlative responsibilities.

Nevertheless, UNCLOS provides a set of international standards and a legal framework to support national and regional authority for protecting both territorial waters and the high seas as well as for managing fisheries on a sustainable basis. The twenty-one years of negotiations and signing were not wasted in that the negotiations prepared the entire industry for a global regime of more responsible management of the ocean commons. The Living Resources section of UNCLOS authorizes coastal states to establish and enforce fishery regulations within their exclusive economic zones. Bolstered a year later by the 1995 UN Agreement relating to the Conservation and Management of Straddling Fish Stocks and Highly Migratory Fish Stocks, known more simply as the Fish Stocks Agreement, UNCLOS requires that states cooperate in conserving migratory fish that move beyond the jurisdiction of their national waters, thereby specifically extending to commercial fish the shared, multinational responsibility for the protection of migratory species granted by the 1979 Convention for Migratory Species. The authority UNCLOS grants to coastal states can potentially benefit entire ecosystems, depending on a nation's environmental and political will and its freedom from economic dependence on fish.

A number of subsequent agreements and failed efforts to craft agreements have worked to tighten the standards and laws and close the gaping holes in the framework that UNCLOS laid out. The UN Convention on Conditions for Registration of Ships that was negotiated as part of the 1986 UN Conference on Trade and Development had a potential for establishing the legal responsibilities of flag states, but it came to nothing. Flag states with large tonnages inserted entry-into-force provisions that gave them the power to render the agreement ineffective, which they did. A 1993 FAO Compliance Agreement attempted to remedy this failure and to close the gap left by UNCLOS between flagging as a right and flagging as a responsibility by calling for compliance with international and regional laws: a state could not flag vessels unless it fulfilled its obligations to the agreement. The fundamental weakness in this agreement, and in all agreements having to do with the high seas, is that it cannot be enforced on the high seas against nonsignatory states, no matter what those states do in violation of the agreements and laws. But an even greater weakness, one that cripples it, is that in ten years, barely enough nations have signed the Compliance Agreement for it to be effective.[5]

The UN Law of the Sea has depended on subsidiary conventions, especially in regard to multinational cooperation in harvesting migratory fish species that cross the boundaries of coastal states into other states and into the high seas. The FAO, with its mandate to improve nutrition on a global basis through crop diversification, reforestation, and fishery management, has fostered some of the subsidiary conventions. Because marine foods are one of its basic interests, over the sixty years of its existence FAO has provided leadership in regard to fish, with increasing relevance to the albatross. Most important in this regard has been the work its Committee on Fisheries (COFI) performed in promoting and sponsoring the 1995 Fish Stocks Agreement and Code of Conduct for Responsible Fisheries, which established principles, standards, and procedures for legally harvesting and transporting fish stocks to markets. Out of the Code of Conduct came two FAO international plans of action that even more directly protected seabirds. As blueprints, the FAO International Plan of Action—Seabirds and the International Plan of Action—Illegal Fishing call upon nations to draw up action plans for conserving seabirds and reducing their incidental capture by both legal and illegal fishers as well as for stopping illegal fishing by their flagged ships.

Established back in 1965 when large-scale commercial fishing was in its early stages, the Committee on Fisheries constitutes the only global intergovernmental forum for addressing fishery issues. The committee examines major international fisheries and aquaculture problems and periodically recommends improvements to fishers and fishery management bodies and to governments and nongovernmental organizations, as well as to the FAO and other international organizations. It also serves as a forum for negotiating global agreements regarding commercial fish.

The Law of the Sea, which went into force in 1994, adopted an ecosystem approach to fishery management and incorporated the concepts of protected areas, pollution controls, and scientific stock assessment. The Fish Stocks Agreement defined the specific obligations of member nations to ensure sustainable, ecologically and socially responsible fisheries. It required coastal states, already empowered in their exclusive economic zones, to influence and to cooperate and comply with fishery management bodies in adjacent waters where their nations' straddling and migratory fish stocks swim. It established twelve governing principles and formally adopted the precautionary approach in regard to stock management. It also accepted that in regard to the high seas, neither flag states nor a global regime would be able to effectively develop, implement, and enforce regulations and laws governing fishing. It therefore designated and authorized fishery management bodies to implement and enforce its governing principles in their jurisdictions, at least in regard to states signatory to the Fish Stocks Agreement. The FSA entered into force in 2003, after thirty states had signed and ratified the agreement, and within three years, it had fifty-six signatories.

The 2003 OECD Round Table paper acknowledges the impossibility of global environmental management, arguing that "effective management can only be done in more tightly prescribed zones and between the most interested parties."[6] The weaknesses some see in the Fish Stocks Agreement are that although nations and fishery management bodies have jurisdiction over migratory and straddling stocks in their waters, the fishery management bodies do not cover all high seas stocks and until now have paid only lip service to the ecosystem approach in management. There is also the problem of ships flying flags of convenience. But the FSA provides a legal enforcement framework, albeit limited to states that have signed it, for fulfilling the Law of the Sea's call for interstate cooperation in managing marine resources. The Round Table Report gave the Fish Stocks Agreement strong praise, saying "this is as close as international law gets to curbing the otherwise sacrosanct rights of flag states and, at least in this respect, [the agreement] starts to place illegal fishing on a par with such obnoxious activities as slavery and piracy."[7] Powerful and potentially far-reaching, the Fish Stocks Agreement stipulates that vessels flagged by a signatory state must comply with all regional and subregional fishery management measures even if the flagging state is not a party to the fishery body in whose waters its vessels are fishing. The agreement also endorses boarding and inspection by fishery officials of a vessel flying the flag of a signatory state on the high seas. Officials must inform the flag state of any serious violation, and if the state fails to take action, the inspector may divert the vessel to port. This agreement of course leaves states that have not signed the FSA untouched, including ones that offend most grievously. But most people see it as an influential agreement and a good step toward empowering fishery management bodies. Whether these fishery bodies are able to take advantage of such empowerment is a different matter. Consensus decision making, lack of enforcement will, and limited resources hamper their effectiveness.

The 1995 International Code of Conduct for Responsible Fisheries also began as a 1991 COFI recommendation. That recommendation led to consultations among UN agencies and relevant national and international organizations to develop a code that was consistent with the Law of the Sea and the Fish Stocks Agreement, which was being developed simultaneously. More than 170 FAO members adopted the comprehensive code. It covers duties specific to states, developing countries, fishery bodies, flag states, and port states to ensure best practices in fishing. The code articulates a global consensus about an entire range of fisheries issues, and urges governments to incorporate its goals and principles into their plans and policies so they can work together, along with fishery management bodies, to accomplish sustainable fisheries. Through the code, the FAO offers to assist nations in expanding and managing their capacities, including aquaculture, setting up habitat protection in coastal zones, and developing methods for responsibly transporting and marketing fish products. A major aim is transparency in the formulation and implementation of fishery management based on the best science. It calls on members to draft enforceable fishing laws, design systems to track fish from the sea to the dining table, and prosecute violators. Because it defines the duties of flag states and provides standards for flagging nations, it also provides support and a framework for efforts to address illegal fishing, especially relevant to CCAMLR's efforts to reduce illegal fishing for Patagonian Toothfish in its waters. It also asks member port states to ensure that vessels allowed into their ports meet national and international commercial fishing standards. The problem is that, as with CCAMLR's conservation measures, compliance with the Code of Conduct is voluntary and required only of signatory nations. As with all soft law, noncompliance is a matter only of embarrassment or shame. On the other hand, any success is an occasion to publicize one's responsible participation in the international community. Importantly for our purposes, the code expresses the goal of reducing wastage, especially among nontarget species, and calls for flag nations to require the use of bycatch mitigation devices on their vessels.

From a 2004 survey of sixty-four member respondents, the FAO found that parties are implementing a mix of measures applicable to their particular situations, be they port, flag, or coastal states. They also noted that there are continuing needs for cooperation on bilateral and regional levels, especially in dealing with illegal fishing. Member awareness of illegal fishing was high, and one-fourth of the respondents were actively drawing up plans to curtail it.[8]

The Code of Conduct is the foundational document for two additional soft-law agreements: the 1998 FAO International Plan of Action for Reducing

Incidental Catch of Seabirds in Longline Fisheries (IPOA—Seabirds) and the 2001 FAO International Plan of Action to Prevent, Deter and Eliminate Illegal, Unreported and Unregulated Fishing (IPOA—IUU). These conference agreements describe the best practices for flag states and call for governments to develop national plans of action to conserve seabirds and reduce losses in marine fisheries. The soft-law provisions of the Code of Conduct and these IPOAs are in effect guidance instruments to assist fishers, states, and regional fishery bodies in accomplishing sustainable marine resource management. Enforcement is replaced by persuasion and encouragement, which highlights the role of national, regional, and nongovernmental leadership and effort in enlisting the voluntary compliance of flag states and the fishing industry.

States have authority over their ports, and vessels must accept the laws of the state whose port they wish to enter. Under the Fish Stocks Agreement, states may pass laws that, if enforced, would impose inspections and prohibit the landing of illegal catches. But this is irrelevant to states that have no concern for stopping illegal landings. Moreover, a fishing vessel that encounters resistance to landing its highly valuable illegal catch can haul up and go to a friendlier port.

In the same year that the Fish Stocks Agreement and the Code of Conduct for Responsible Fisheries were signed, two men, thinking independently of each other, came up with the idea of an environmental certification program for the trade in wild fish. Anthony Burgmans was then chairman of Unilever, a global seafood wholesale business that began in 1930 as a conglomerate of nineteenth-century soap and food wholesalers. Dutch-born Burgmans joined the firm in 1972 as marketing assistant and rose through the ranks to guide one of today's largest companies. With a liking for wildlife, Burgmans recognized in the 1990s that fish were becoming more important for nutrition worldwide, but he also saw that for Unilever's frozen fish business to survive, something had to happen to conserve diminishing fish stocks. He felt that Unilever alone, despite its significant presence in the market, could do little to resolve the problem of collapsing fish stocks. He decided "we had to work with others to leverage change"[9] and found Mike Sutton, who was leading the World Wildlife Fund, now the World Wide Fund for Nature (WWF), who was thinking along the same lines. In 1996, Burgmans and Sutton began talking about how to assure sustainability of fish stocks and the integrity of the marine ecosystem worldwide. The idea was denounced from both sides, by conservation nongovernmental organizations and by the fishing industry, and Burgmans says, "It was a lonely road when we and WWF started."[10] Moreover, they encountered the same dilemma that fishery management bodies continually struggle with: determining how to establish policy in a context of inadequate and conflicting science and powerful political and economic interests. No one to date agrees on how to verify that a fishery is technically sustainable, especially since fish are migratory, natural and climatic conditions are constantly changing, data are imprecise and limited, and illegal fishing results in a significant understatement of catch data.

Nevertheless, by 1997 the Marine Stewardship Council was engaged in a global consultation process with stakeholders from around the world to establish a set of standards for sustainable fisheries based on the 1995 UN Code of Conduct for Responsible Fisheries. The outcome of this hard work was the continually evolving set of principles and criteria the MSC uses to identify a fishery as sustainably fished and to label catch from that fishery as certified along the chain, from the delivery port to wholesale and retail markets and, ultimately, to the table, fulfilling Burgmans's goal of "making it harder for anyone to sell unsustainably harvested seafood" and moving one step closer to changing "the whole way a market works."[11]

In 1999, the Marine Stewardship Council became a fully independent nonprofit organization and developed over subsequent years a governance structure that incorporated a balance of views, representation from all sectors, and increasing transparency of its operations. It established a stakeholders' council that meets annually to draw up recommendations that the main council board must consider. Additionally, the two joint chairs of the stakeholders' council have seats on the main council board. As the current chief executive Rupert Howes notes, "We've spent millions of dollars improving our governance and procedures. We had genuine issues with nongovernmental organizations and there were real concerns about quality and consistency of certification."[12] FAO standards require that any certification entity for food products such as agriculture and fish must be independent, science-based, and inclusive of all opinions on the criteria for the certification. Achieving this level of quality took the MSC ten years. Independent parties now conduct the certification process following council criteria. To be certified as sustainable according to these criteria, a fishery must score 80 percent or more in each of the following three standards:[13]

· The fishery is conducted in a way that does not lead to overfishing or depletion and demonstrably leads to recovery of depleted stocks.
· Operations allow for maintenance of structure, productivity, function, and diversity of the ecosystem on which the fishery depends.
· The fishery is subject to an effective management system that respects local, national, and international laws and standards, and incorporates institutional and operational frameworks that require responsible and sustainable use of the resource.

In 2000, the MSC certified its first fishery, Western Australian Rock Lobster. In 2003 it certified six more fisheries, and in 2004 added four more. By the close of 2005, there were fourteen certified fisheries and over 300 seafood products labeled with the MSC's blue fish logo for sale in supermarkets around the world, and Howes reported in 2006 that 30 percent of prime whitefish worldwide was in the certification program. In that year, the global sale of £100 million in MSC-certified fish reflected an increase of 16 percent in one year alone. In 2007, the MSC listed twenty-two certified fisheries, which in the Southern Ocean include Australia mackerel, New Zealand hoki, Patagonian scallop, South African hake, and South Georgia Patagonian Toothfish. It had an additional fifteen fisheries undergoing certification, some thirty under assessment, and more than 450 products for sale. Three years later the number of certified fisheries had risen past one hundred. It has garnered a broad coalition from one hundred or more support groups in approximately twenty nations since it began to operate independently. As hoped for, some fishery management bodies and fishing operations are improving their target fishery management so they can participate in the MSC program.[14]

The South Georgia Patagonian Toothfish fishery is an example where fishing practices improved in order to meet MSC criteria. However, this certification has not occurred without a great outcry. Because of the high price Chilean Sea Bass commands, this fishery, under British control but in compliance with CCAMLR regulations, is a prime area for illegal fishing. The MSC's March 2004 certification of a total allowable catch in that fishery of 4,230 tons of toothfish taken demersally between May and August caused dismay in some bird and fish circles. Opponents of certification of the South Georgia fishery referred to actions that had been taken against vessels practicing illegal fishing in and around South Georgia and claimed that the fishery was an isolated, limited toothfish population that should not be exploited. The South Georgia fishery harvested less than one quarter of the fifteen thousand metric tons CCAMLR allowed at the time, but the MSC recognized that the illegal catch was probably five or six times larger than the legal limit. Despite the objections, the final report and decision of the five-member objections panel recommended certification. It refused to use the likelihood of illegal fishing in the area as a reason for not certifying legally caught fish, remarking that it was not MSC's duty to certify the entire toothfish industry. In September 2009, the fishery received recertification.[15]

Some regard the MSC as compromised by this certification. Toothfish are currently a lightning rod for seabird bycatch, overfishing, and illegal fishing. But conservationists point out that the South Georgia fishery now has independent observers on all its boats, is closed for six months of the year, and is reckoned to be the best-managed longline fishery in the world. It is not MSC certification alone that has strengthened management of the South Georgia fishery, however. Operating undercover and using sovereignty and de facto rights, the United Kingdom has invested millions in surveillance to sort out and expel illegal fishing. Unfortunately, as always occurs with effective efforts to stop illegal fishing, the vessels that were fishing surreptitiously moved illegal harvesting into the Indian Ocean, pointing out the limitations of enforcement that is less than global.

It was reportedly an impressive sight to see five ocean-hardened longliners tied up in Stanley, Falkland's capital, in early September 2007, as they waited for certification inspection by the Marine Stewardship Council. Each vessel carried an onboard observer and had participated in a live fish-tagging program that tagged and released three fish for every two tons harvested. Of the more than 5,000 toothfish tagged, 583 of them have been recaught. The plan was to monitor the tagged fish to see where they swam and how many were recaptured as a way of determining more accurately sustainable quotas. Longliners fishing in the South Georgia fishery are required to set their lines under deterrent streamers and at night, and the known albatross and petrel bycatch is reportedly minimal.

Despite setbacks and naysayers, the Marine Stewardship Council appears to be having positive effects in some fisheries. John Gruver of United Catcher Boats in the United States, a consortium of small fishing vessels, said that in the absence of a race to fish, fishermen are now competitive about bycatch avoidance. Large bycatch of salmon has resulted in the adoption of a rule that when the salmon in a boat's bycatch reaches the official salmon limit, the boat stops fishing even if it hasn't caught its limit of the target fish. "The fisherman of the future isn't going to be measured by the fish he does catch, but by the fish he doesn't catch," Gruver adds.[16] The Marine Stewardship Council's Howes doesn't attribute full responsibility to the MSC for all these improvements but does believe that it has made a significant contribution to them in the form of standards and incentives.[17]

The council also gives its MSC certification and label to wholesalers, retailers, and restaurants that deal only in MSC-certified fish, and this is producing additional benefits by further increasing consumer power to affect the market. MSC asks its certified suppliers to confirm the legality of marine products they are buying and selling. WalMart has adopted a goal of selling 100 percent MSC-certified fish in the United States within five years. Whole Foods sells only MSC-certified fish, and other major supermarket chains are moving toward complete participation, including ASDA, Compass, Metro, Safeway, Marks & Spencer,

Waitrose, Sainsbury's, and Tesco. McDonald's in the United States uses MSC-certified fish but receives little credit for this practice because it prefers not to pay royalties for the use of the MSC label.[18]

Early in 2006, the MSC conducted a survey of the leading UK supermarkets and was pleased that they were increasing efforts to procure fish from sustainable supplies. The council survey placed Marks & Spencer first for having the best practices designed to ensure that all their fish came from sustainable sources. In response to the chain's intention, Rupert Howes, MSC's chief executive, stated, "This is a very welcome commitment from a national retailer and underlines Marks & Spencer's long-standing efforts to offer green products and to respond to the environmental concerns of the public."[19] MSC discovered that Marks & Spencer was selling fifteen species from the MSC's "Fish to Eat" list and only one from the "Fish to Avoid" list. Another major outlet, Sainsbury's, noted that its consumers, too, were becoming more aware of population declines and wished to avoid purchasing endangered fish. The MSC encouraged supermarkets to label not only certified fish but also those that are threatened or endangered.[20]

In addition to royalty income, the Marine Stewardship Council is funded 80 percent by trusts and foundations, primarily in the United Kingdom and the United States but also in the Netherlands and Sweden, and by nongovernmental organizations such as ComMark Trust, World Wildlife Fund International, and the World Wildlife Fund US; 3 percent by the UK's Department for Environment, Food and Rural Affairs and Sweden's International Development Agency; and 6 percent by five corporations, the Fish Society, Marks & Spencer, Tesco, Unilever, and Whole Foods Market.[21]

The effectiveness of certification in protecting fish depends on the integrity and transparency of the process. The Marine Stewardship Council took a hit in late 2007, when despite objections from New Zealand's Royal Forest and Bird Protection Society and the World Wildlife Fund, they recertified New Zealand's hoki fishery. Those opposed claimed that the fishery lacked a recovery plan despite the 60 percent decline in catches since the fishery was first certified in 2001 and that it was killing seabirds and marine animals and damaging the seabed. Forest and Bird spokesman Kevin Hackwell called the certification process fatally flawed and argued that it sent the wrong message to commercial fishers worldwide. The independent objections panel reported that the numerical score awarded by the assessment body passed by the "skin of its teeth." The panel found that the certification body's decisions were overly optimistic in regard to the assumptions they used about fish recruitment. Using different assumptions, the panel found that New Zealand's hoki fishery was at best barely sustainable, and more likely unsustainable. This was a severe challenge to MSC credibility, and the organization was forced to acknowledge unprofessional decision making and change its operating policies.[22]

Recently the MSC has received flak because of its certification of a krill fishery operated under CCAMLR regulations that permits a limited harvest designed to accommodate natural predation by seabirds, fur seals, and whales in one specific area of CCAMLR jurisdiction. The Norwegian company Aker BioMarine received the MSC logo in May 2010, granted for eighty-six fisheries with some five thousand sea-based items worldwide, including krill-oil pills. But again, as with the hoki fishery, some marine scientists and interested officials believe the MSC is compromising its certification system through the use of third-party assessors who have an interest in commercial outcomes. Southern Ocean groups have strenuously objected to the MSC certification in connection with krill, the major support of Antarctic marine life, and another reportedly questionable certification of toothfish, sold as Chilean Sea Bass and harvested in Antarctica's Ross Sea.[23]

Nevertheless, today the London-based Marine Stewardship Council is widespread in its operations, with offices also in Berlin, Cape Town, Edinburgh, Seattle, Sydney, The Hague, and Tokyo. Its mission is "to safeguard the world's seafood supply by promoting the best environmental choice."[24] The mechanism by which it fulfills its mission is to provide the information the individual consumer needs to shape the market according to sound environmental principles. When individual consumer power achieves critical mass, it will shift the market-based incentive structure for commercial fishers, who will unhesitatingly change their practices to maintain profitability. At least this is the dream. The problems of uninformed and unmotivated consumers have slowed but not defeated the realization of the great potential that resides in the Marine Stewardship Council and consumer power.

Another important piece of the picture for saving the albatross that emerged in the mid-1990s was a recommendation in 1997 at the twenty-second meeting of the FAO's Committee on Fisheries to obtain expert consultation, using extrabudgetary funds, on reducing seabird bycatch in longline fishing. The consultation would provide the basis for an international plan of action for reducing seabird bycatch. The United States and Japan agreed to fund the consultation. FAO set up a Seabird Technical Working Group of eighteen members from various governments and nongovernmental organizations and scientists closely involved with albatross research. For background, the Working Group reviewed three studies of seabird bycatch and mitigation in longlining and combined them into one document it published in 1999 as an FAO Technical Paper, *The Incidental Catch of Seabirds by Longline Fisheries: Worldwide Review and Technical Guidelines for Mitigation*. The publication compiled research by Nigel Brothers on seabird mortality in tuna longlining, by Norway's marine specialist Svein Lokkeborg, and by South Africa–based John Cooper, the first director of BirdLife International's 1997 Global Seabirds Program. The paper described the world's major longline fishing operations and provided the most current numbers on the incidental catch of seabirds. It also summarized best-practice mitigation measures and their cost-effectiveness and provided technical guidelines for their application. Mitigation measures included nontechnical approaches such as seasonal fishery closures, education and training of members of the fishing industry, and direct efforts to increase seabird populations. The report noted that lack of education was the most frequently mentioned impediment to effective mitigation, followed by noncompliance with national and regional fishery management bodies in the use of mitigation devices and procedures. Just as the World Albatross Conferences had gathered and then disseminated the current knowledge about the albatross family, the FAO's COFI report publicized worldwide the available technology and training for seabird bycatch mitigation procedures as well as the importance of using them.

Latin American members responded to this paper with criticism for the Seabird Technical Working Group's failure to include all regions in its study and to recognize and acknowledge that incidental losses were more serious in some fisheries than others. The problem was minimal in tropical seas, for example. South American scientists called for more research and also argued that funding was needed for the development and implementation of any plan of action for seabirds. Others argued that bycatch reduction required a plan of action to be adopted on a comprehensive, not regional, basis, and that the plan should be broad enough to be implemented in all fisheries. Noting that input, in October 1998, the official FAO consultation on seabird mortality took place, on the basis of which COFI developed an International Plan of Action for Reducing the Incidental Catch of Seabirds in Longline Fisheries (IPOA—Seabirds), published in 1999. The plan lists tuna, swordfish, and billfish fisheries as places where seabirds are taken in longlining, especially albatrosses and petrels in the Southern Ocean and in the tuna fisheries to the north of CCAMLR's boundary. The plan also recognized the danger to seabirds from longlines fishing for toothfish and noted that bycatch in unregulated toothfish fishing was significantly higher than that in CCAMLR's regulated toothfish fishery. As many as 145,000 seabirds were lost in the toothfish fishery in the 1996–1997 season.[25]

The International Plan of Action—Seabirds provides comprehensive information about devices that reduce seabird casualties. Implementing them is voluntary, however. States are to determine whether there is a problem regarding seabird bycatch, and if so, FAO experts may offer funding and technical assistance to them in drawing up their national plans of action for reducing seabird bycatch (NPOA—Seabirds). IPOA—Seabirds, in the development of which BirdLife International and Australia took lead roles, had an ambitious schedule. It called for fishing nations to set up and begin implementing their own NPOAs within two years, with review and revisions thereafter at four-year intervals. The FAO serves as the depository for biennial reporting on the implementation process.

The assessment and reporting provisions of the IPOA—Seabirds call for monitoring and systematic data collection to determine the type and extent of seabird bycatch. Data include fleet numbers and vessel sizes in a specific area, methods of fishing, fishing areas and effort, seabird populations in those areas, numbers of known bird losses, and mitigation measures in place. Ten recommended technical measures follow CCAMLR guidelines, including increasing bait sink rates, underwater setting, use of bird-scaring lines, use of bait-casting machines, and additional deterrents used during line setting. Five operational measures focus on setting lines at night, controlling discards, releasing live birds, closing areas to fishing when breeding birds are abundant, and, most importantly, licensing fleets that use mitigation as standard practice. IPOA—Seabirds is a soft-law action plan, elaborated under a voluntary framework similar to that of the Code of Conduct for Responsible Fisheries, which establishes standards of behavior "with a view to ensuring the effective conservation, management and development of living marine resources with due respect for the ecosystem and biodiversity."[26] Both FAO documents have proven to be valuable templates for nations in formulating their respective plans of action and best practices, and FAO's funding and technical assistance have been essential. In 2001, FAO followed its International Plan of Action for Seabirds with an International Plan of Action for stopping illegal fishing, also essential to saving the albatross.

After the expert consultation FAO organized in 1997 and meetings among COFI members that resulted in IPOA—Seabirds, the widely hailed voluntary seabird action plan drew a lukewarm response in respect to implementation. Only nine nations published action plans during the first decade of its existence, and experts found disheartening differences in consistency, content, and effectiveness among national plans. Consequently, in 2008, FAO held another consultation in Bergen, Norway, aimed at streamlining, and an updated IPOA—Seabirds recommended that fishers explore a suite of mitigation measures for longlining that had proven effective for seabird avoidance. The consultation also sought to galvanize states and fishery management organizations into increased cooperation and transparency in collecting and sharing data about bycatch and to institute and expand an at-sea observer program in their fisheries in order to implement and enforce rules of best practice. Importantly, the expert consultation, composed of nine fishery management bodies seabird experts, and two resource advisers, moved the IPOA—Seabirds in a new direction to include the effects of trawl and gill nets on seabirds, a disturbingly chronic problem that fishers and bird experts are exploring. The review concluded by encouraging states and fishery management bodies to study the effectiveness of conservation actions at both the vessel and fishery levels, and to ensure practical and relevant onboard training and periodic assessment of changes in seabird status in the fisheries.[27]

On another front, just as the FAO was planning in 1997 to obtain the consultation on seabird bycatch mortality and mitigation, BirdLife International launched an ambitious program with funding from its British partner the Royal Society for the Protection of Birds. BirdLife started a Global Seabirds Conservation Program, drawing on the solid global science available to it as well as the strength and reach of its organizational base. BirdLife International is a global alliance of conservation organizations, each responsible for a particular territory or region, in more than one hundred countries. There are partners, partner designates, and affiliates, each with different but overlapping autonomy within home regions and territories. Currently there are partners in about 115 nations in seven regions, including the islands around Antarctica, and this number continues to rise. Staff conduct research and analyze the status of birds, bird habitats, and the problems threatening the world's ten thousand or so species. The alliance identifies major environmental issues pertaining to bird life and influences the way nations, international organizations, and other nongovernmental organizations set priorities for environmental improvement. It evolved from a smaller organization, the International Council for Bird Preservation, established in 1922 by T. Gilbert Pearson, president of the US National Association of Audubon Societies, who wanted to unite US and British bird-protection efforts in the same way they had come together to fight the fashion plumage industry in 1910. Headquartered in Cambridge, England, BirdLife today has a staff of four thousand, more than 2.5 million members worldwide, and ten million supporters. It owns or manages reserves totaling more than 3,800 square miles. It holds a global partnership meeting every four years to formulate policies and negotiate strategies for implementing research and conservation programs. It elects a global council and regional committees, and the council, in turn, appoints a chief executive to head the international secretariat that coordinates and supports the world partnership.[28]

Global Seabirds, initially run from South Africa's Cape Town under John Cooper and later Deon Nel, was a long-range strategic plan of operations for seabird conservation at all levels—national, regional, and global. Global Seabirds recruits and collaborates with partners to formulate seabird protection policy on a worldwide oceans scale. Its work and informed advocacy also contributed to the construction of the FAO's 1999 International Plan of Action—Seabirds. Global Seabirds collaborates with fishers worldwide to identify the mitigation equipment and procedures best suited to specific fisheries, conducts training workshops, and places observers on vessels. Eventually, the program led to the formation of the 2001 hard-law Agreement for the Conservation of Albatrosses and Petrels (ACAP). In 2004, the program's leadership shifted to Sandy, Bedfordshire, home of the Royal Society for the Protection of Birds, where a three-member team works for seabird protection with special attention to longline fishing. Since 2006, this Albatross Task Force has targeted areas of major bird losses, notably in Latin America and southern Africa, and recruited local workers to observe bycatch problems aboard vessels and negotiate with fishers and officials about the best practices for reducing seabird losses.

By 1999, most of the major pieces for dealing with the long-standing mess of the high seas and for protecting seabirds were in place, with the exception of the 2001 Agreement for the Protection of Albatrosses and Petrels. The big picture was emerging, and one could see the outlines and building materials of an ethical, political, and economic network of international standards, regulations, and soft laws reaching to connect, some starting at the bottom and building up, others coming together at the top and building out and down. Many pieces were still missing, and many still are. Everyone who plays a role in this story had to work to some extent on blind faith, just as Robert Cushman Murphy and his early colleagues in Southern Ocean bird science did, having no idea where their work would lead or the role it would play in any larger picture. Many of the players have been politically adept, extremely so in some instances. They know how to think and operate strategically, selecting the time and place and specific person or group to mobilize, antagonize, include, or exclude. As John Croxall says, in working to get at the underlying truths and craft solutions to the problem, "it's been partly serendipity, partly crafted Machiavellianism, and really good people thinking together."[29] In working with national governments, with commercial fishers, with well-meaning but obfuscating bureaucrats, it's been important to frame the argument in terms of what matters most at the moment. If using mitigation devices saves fuel in the long term, talk about bycatch mitigation in terms of fuel savings to fishers who are concerned about the cost of fuel. If an agency has a conservation commission but lacks adequate science and data, mobilize efforts to fill in that gap without taking away the buy-in or intruding into the operations of the agency. If Greenpeace is effectively playing its role of policing and demanding extreme action, use the demands strategically to create the space and political will for achieving more modest gains. Every stakeholder of any persuasion has a role to play, and organizations like Greenpeace and the International Fund for Animal Welfare are strategically essential.

Now it remains to see how the major players with their emerging pieces are playing out the story, given that some pieces are still missing. What began here as a disconnected story of lone individuals working against all odds to follow their passion, a story that banded albatrosses—or color-marked or transponder-bearing albatrosses—bird scientists, and passionate observers had to piece together well into the 1980s, evolved into a story of emerging structures that overlapped and interconnected to begin to form an environmental regime for the oceans. Our challenge changes from that of piecing together the story in terms of individuals to that of trying to describe how international standards and agreements, fishery management bodies, nations, nongovernmental organizations, commercial fishers, and celebrities have interacted to build, maintain, strengthen, and extend an environmental regime for the albatross. Chronologically, we've come a long way from the Moriori's selective harvesting of albatrosses on the Chathams to ensure their survival and from Robert Cushman Murphy's early scientific observations of the albatross's interactions with whaling in its harsh Southern Ocean habitat on South Georgia. Events are now taking place in the highly technical and international context of a competitive global market economy, but we're still trying to ensure the survival of the albatross. In that sense, we've returned to the beginning of the story, to the Moriori and their selective harvesting of albatrosses to ensure repopulation. It reminds us of James Joyce's long book *Ulysses*, which closes with the observation that the "longest way round is the shortest way home." We've definitely taken the longest way round in regard to our relationship with living resources. Are we any closer to home?

Part Five

AGENTS OF CHANGE

CHAPTER 17

Fishers

A WEAKNESS IN FISHERY management since the first ocean management bodies were organized has been their dependence on the will of national members to support and enforce the regulations upon which the nations themselves have agreed. This has certainly been the case with the fishery management bodies with significant numbers of albatrosses in their waters. But it also seems certain that, because of the internal resistance from powerful economic interests, nations would not have been able to address issues of fish stocks sustainability or bycatch on their own, without the pressures and models provided by the regional fishery entities.

When CCAMLR membership grew concerned about Patagonian Toothfish stocks, especially as they recognized the extent of illegal fishing, CCAMLR added the fish to its list of regulated species and began requiring a permit to harvest Patagonian Toothfish in CCAMLR waters. But it wasn't until the late 1990s that CCAMLR formally began addressing illegal fishing. When its Scientific Committee questioned the sustainability of toothfish stocks, CCAMLR's Standing Committee on Implementation and Compliance worked out two schemes that placed CCAMLR in a leadership role in regard to monitoring legal and policing illegal fishing.[1] It initiated in 1999 a catch-documentation scheme for Patagonian Toothfish to give fishers accountability and their catches transparency (see Appendix). Shipmasters had to track and document their catches from the water to the primary market, providing the name, homeport, registry, and call sign for their vessels. They also had to record the permit number, weight, place, and date of each catch and were bound to follow it, through transshipment at sea if necessary, to the point of landing. Additionally, each flag state had to certify the documentation of its flagged vessels. An additional measure agreed upon in 2000 required each party's vessels, under threat of prosecution for making a false statement, to notify port authorities in writing prior to arrival to offload a catch.

Members subsequently resolved to deflect from their ports the flag vessels of nations known not to be implementing catch documentation and to avoid flagging any vessel proven to have engaged in illegal fishing. They further resolved not to allow vessels owned by their nations' firms to be flagged to nonmember states. Despite some members' failure to respond in good faith to these resolutions, the resolutions themselves, reached through consensus vote, were a significant step forward. In recent years, a method of blacklisting parties and vessels for illegal fishing has amplified the scope of CCAMLR from a focus solely on operations within its area to include illegal operators in nearby waters. Standardization of documentation forms; inclusion of Lloyd's registration numbers for vessels, which, like the *Viarsa 1*, as we shall see, frequently may switch names; and new codes for classifying toothfish in trade statistics are recent efforts toward greater transparency. The recent decision by Mauritius, a chronic landing port of convenience for illegal fish, to enforce catch-documentation requirements is also a move in the right direction.

In 2000, CCAMLR also began requiring all vessels licensed to take toothfish in its area to carry geographical positioning systems (GPS) to allow monitoring of the location, date, and time of their activities. It's not a tamper-proof system. CCAMLR's Standing Committee on Observation and Inspection reported in 2002 that two Uruguayan-flagged vessels were spotted at least one thousand miles south of where their vessel monitoring system and catch documentation positioned them. Uruguay, a CCAMLR member, imposed no penalties on these operators, but the government management authority was charged with collusion and reportedly has now cleaned its house. Japan, Poland, South Korea, and Ukraine objected to being required to use monitoring devices on krill vessels and unilaterally exempted themselves from the requirement, providing their unscrupulous fishers a free rein. Most recently, CCAMLR discussions have focused on upgrading vessel-monitoring systems, including harmonizing shipping codes for toothfish. In adopting these two systems and resolutions, CCAMLR has more clearly and forcefully aligned itself with the best practices as laid out by the Fish Stocks Agreement, especially as they pertain to stocks that move beyond the sovereignty of a coastal state.

The requirement for consensus-based decisions, the incompleteness of data and knowledge, a dearth of members' political will for compliance, and the ubiquity of illegal fishing remain serious obstacles to CCAMLR's effectiveness.

More knowledge is an immediate and urgent need. In his testimony to a 2003 select committee of the UK House of Lords devoted to science and international agreements in polar regions, John Croxall emphasized the importance of fostering scientists and organizations committed to delivering applied science into environmental management, even though "it is not seen as something for which they get much credit in terms of their career development."[2] He lamented, in reference to CCAMLR, that "of the 24 full members, probably fewer than half a dozen are engaged at all levels in the provision of scientific data, analysis, advice and discussion."[3] This is much more the case with the other fishery management bodies with significant albatross presence in their areas.

Comprehensive management and policing of toothfish fishing and trade are now imperative. Patagonian Toothfish is the most valuable fish currently harvested in CCAMLR waters and thus a prime target of illegal fishing. Neither CCAMLR membership nor its measures and quotas have prevented illegal operators from maneuvering inside area waters or from operating within national exclusive economic zones, such as HIMI and Kerguelen, within or adjacent to CCAMLR's area. At the turn of this century, Traffic Network, an international organization that monitors wildlife trade, reported that fifty-six nations were engaged in the Patagonian Toothfish industry, many of them with more than sustainable management in mind. It charged ten of twenty-four CCAMLR members with having recently engaged in illegal Patagonian Toothfish fishing: Argentina, Spain, the United Kingdom, Uruguay, Belize, Denmark, Panama, São Tomé and Príncipe, Seychelles, and Vanuatu. CCAMLR had asked eight of them to cooperate with its catch-documentation scheme, and the remainder may not have known about regulations to limit toothfish catches. But it is likely that fishers from some of those nations were knowingly fishing illegally.[4]

Negative global publicity and embarrassment within the membership is about all that can be used at this point. That some CCAMLR members have convenience-flagged some miscreant longlining and trawling vessels and resisted efforts to toughen regulations about illegal fishing causes tensions. As the annual legal toothfish take fell two-thirds between the 1996–1997 and 1999–2000 seasons, experts and officials drew comfort from the extra protection and surveillance CCAMLR was providing. But at the time, Traffic Network estimated that illegal operators were in fact harvesting four times the lowered legal limit and reported that these fishers were extremely resourceful in bringing illegal catches to markets.[5]

Greenpeace has called for a total moratorium on Patagonian Toothfish capture, arguing that without a complete ban, illegal fishers will still fly flags of convenience, falsify catch statistics, and continue to catch and unload toothfish.

Although CCAMLR rejected this proposal, CCAMLR members Australia and the United States recently attempted to have Patagonian Toothfish classified as an endangered species by the 1975 Convention on International Trade in Endangered Species of Wild Fauna and Flora (CITES). If toothfish are included in the CITES Appendix II that lists vulnerable species for which trade should be controlled, all toothfish fishing would be subject to restrictions like the ones CCAMLR's members are expected to follow. Most importantly, illegal toothfish fishers would no longer be able to use ports that fall under CITES jurisdiction, including ports of the 172 CITES parties.[6] Although CITES parties reported in 2007 a decline in illegal toothfish harvests, they also acknowledged that some CITES members were fishing illegally for Patagonian Toothfish in the CCAMLR area with vessels flagged to Equatorial Guinea and Togo. CITES member Singapore was found to be in only partial compliance with the CCAMLR catch-documentation requirements, and neither Indonesia nor Hong Kong had signed on to the documentation scheme. At the urging of Australia and the United States, CITES parties renewed a 2002 resolution of cooperation between CITES and CCAMLR regarding the toothfish trade. CCAMLR also urged that CITES parties implement CCAMLR's catch-documentation scheme, gather information about illegal fishing, and supply it to the CCAMLR Secretariat in Hobart.[7] But CITES parties took no action to list the two toothfish species in Appendix II. Adding toothfish to the list has support from the CITES Secretariat, but it has not happened to date. Nor has information flowed to CCAMLR from CITES parties regarding illegal fishing. At best, the two conventions are talking to each other. Developing a cooperative and collaborative working relationship between CITES and CCAMLR is today an area for the application of some deft political skills and creative leverage by the various stakeholders.

It is clear that CCAMLR has moved more aggressively in recent meetings to deal with overfishing, illegal fishing, and bycatch. But the pace of movement does not satisfy environmental advocates alarmed by the continued erosion in fish and bird populations. Member nations have been able to block more progressive recommendations and slow down actions that impinge on national interests and economic returns, thereby facilitating the exploitation that continues to threaten commercial fisheries. The problem is greatly exacerbated by long-term and significant overcapitalization of the fishing industry worldwide and the resulting overcapacity of high-seas fleets. Maintaining an adequate return on investment puts great pressures on the industry to harvest immature and fragmented fish populations and even to fish illegally. The clear indications that Patagonian Toothfish populations have been depleted in the all-too-common boom-and-bust cycle of commercial fishing give critics a plausible case for arguing that CCAMLR has done too little, too late, despite its being a leader in fishery management in every respect. However, CCAMLR has stepped ahead of the other fishery management bodies in regard to albatross and other seabird conservation, with significant results.

In 1989, when CCAMLR's Commission formally acknowledged the problem of the incidental capture of seabirds in longline fisheries, it only urged members to voluntarily implement mitigation devices and procedures on their vessels. This itself was a breakthrough, however, given the consensus requirement for formal decisions. Members' understanding of the problem deepened because of the excellent body of scientific data and expertise available to the Commission from its Scientific Committee, but achieving adequate regulation of seabird bycatch and enforcement of bycatch mitigation proved to be an arduous and often discouraging task. As they understood more and more about the dynamics of the ecosystem and how, where, and how many birds were being killed, some members began to call for specific measures backed by enforcement mechanisms.

In sustaining this effort, the leadership shown by scientists like John Croxall, of fishers like Chile's Carlos Moreno and Jorge Berguno and New Zealand's Malcolm McNeill, and of other colleagues in convention delegations has proven critical. Croxall was a founding member of the Scientific Committee and of the 1993 ad hoc Working Group on Incidental Mortality Arising from Longline Fishing (WG-IMALF; see Appendix). He has worked as a scientific adviser, consultant, and diplomat-negotiator on albatross status and protection at all levels. Working as both scientist and advocate in the highly politicized arena of environmental networks and operations, he is uniquely skilled at integrating science into the policy development process for Antarctica and the Southern Ocean. He argues that ensuring an appropriate balance between conservation and rational use requires the best and most comprehensive science available. By all reports, he is also a master at delivering reliable science into the applications and management area. Because of Croxall's credibility and political skill, CCAMLR members were able to hear and respond to his call for action in his 1990 editorial in *Antarctic Science.*

Croxall ran the Working Group on Incidental Mortality for seven years, until 2004. Consisting mostly of bird scientists, the working group reported to the Scientific Committee, on which Croxall continued to play a leading role. Crucially, the working group was first convened by Carlos Moreno, a fisher and adviser to the fisheries delegation from Chile who is also a biologist and ecologist at the Instituto de Ecología y Evolución, Universidad Austral de Chile, based in Valdivia.

Chile's delegation was headed by its respected ambassador Jorge Berguno. Moreno has published an educational manual for fishers and coauthored an article with John Croxall. In the Working Group on Incidental Mortality, the Chilean representatives pressed hard for greater national accountability in regard to seabird bycatch. Because fishers held Moreno in such high esteem, his leading role in the Working Group on Incidental Mortality gave a very powerful message to fishers and fishery management bodies. According to Croxall, his leadership and credibility made it possible for the working group to be heard by some of the more resistant CCAMLR member nations such as Russia and Ukraine. With his scientific grounding and experience as a fisher, Moreno also mobilized fishers throughout South America in the effort to reduce seabird bycatch. In 2006, he was an active partner in a Fishers Forum in São Paulo, Brazil, where sixty-five participants from seven nations learned about best practices and the latest innovations in reducing seabird losses. Fishers and industry representatives, including ship owners, made up almost half the participants. Having headed the first Fishers Forum six years earlier, Moreno recapped what had been accomplished and what challenges lay ahead, notably that of expanding observer coverage.[8]

CCAMLR's Working Group on Incidental Mortality met for the first time in Hobart in October 1994 and took over a significant part of the Scientific Committee's workload, which had been increasing as a result of the concern with seabird deaths. The working group assessed the gravity of incidental mortality and reported to the Scientific Committee, advising it on best practices for convention waters. In turn, the Scientific Committee made recommendations to the Commission. The Commission responded by increasing oversight and accountability for fisheries and fishing issues in three ways, based in large part on the work of people like Nigel Brothers and Graham Robertson. First, starting in the mid-1990s, CCAMLR began to require the use of specific mitigation devices and procedures to reduce mortality among nontarget animals. Second, it mandated that observers be placed aboard longliners to gather hard data about bird deaths on a species level, and in 1999 published a bird guide for fishers that illustrated ways to reduce bird deaths while setting and recovering lines. Third, CCAMLR also delayed the start of the fishing season in some areas in order to protect breeding albatrosses.

A series of binding Conservation Measures mandated that boats adopt six procedures: use of lines designed to sink quickly, night setting, use of bird streamers, discharge of offal free of hooks on the opposite side from where the main line is set and never during line setting, use only of thawed bait, and the live release of captured birds.

These requirements remain the basis for CCAMLR's bycatch mitigation program as well as, to some extent, those of other fishery management bodies that address bycatch. From time to time, complaints surface that fishing masters and crews are not following them, but CCAMLR holds an impressive record for lowering seabird bycatch in the Southern Ocean. In the 2001–2004 fishing seasons, the CCAMLR Secretariat reported a very low take of seabirds in the regulated toothfish fishery, down from a rate in 1997 of 0.5 birds per 1,000 hooks. There were no reported losses in 2002 or 2006 in the CCAMLR area, but the rate increased again in 2007.[9] Although losses spiked in French sub-Antarctic waters, they began to fall when fishers there began adding weights to their lines. The Secretariat believed that the lower rates were the direct outcome of mandatory onboard observers as well as increased compliance and seasonal closures. But the Commission expressed ongoing concern about the extent of illegal fishing in CCAMLR waters. In 2009, unregulated and illegal longline vessels, which don't bother with streamers or other seabird deterrents, reportedly took almost one thousand tons of toothfish in CCAMLR waters. In a survey for Traffic International, which is owned by the World Wildlife Fund (IUCN), at about the same time, Mary Lack estimated that evasion of vessel monitoring and catch documentation resulted in a significant underestimate of illegal landings of toothfish, which were likely a good deal higher than CCAMLR estimates.[10]

CCAMLR ranks lowest in albatross presence in comparison with the five other fishery management bodies with albatrosses breeding and foraging in their waters. Twelve species appear in CCAMLR waters, compared to eighteen in CCSBT, WCPFC, and IOTC (see Table 2) areas. As a result, although CCAMLR is a template for what can be done, its effect on the reduction of albatross deaths is limited by its geographical location, despite its vast reach. The tuna commissions, on the other hand, have major populations of birds in their fisheries, and they are beginning to take steps in response to the Law of the Sea, the Fish Stocks Agreement, and the Code of Conduct for Responsible Fisheries.

The 2004 Western and Central Pacific Fisheries Commission, a product of the 1995 UN Fish Stocks Agreement, also proclaims itself a pioneer among regional fisheries bodies because of its ecosystem and precautionary approaches to fishery management. Its mandate upon emergence in June 2004 was to conserve highly migratory fish stocks, mostly tuna, in the Pacific Ocean. Based in Christchurch, New Zealand, in 2008 the commission had twenty-five members, seven participating territories and seven cooperating nonmembers in a seascape that is one of the most productive fisheries on the planet. It has 150 million square miles under its jurisdiction, spreading from CCAMLR's 60°S latitude boundary

191

northward through Hawaii, Japan, and into the Gulf of Alaska. It begins on a line of longitude west of the North American coastline and takes in islands such as French Polynesia, New Zealand, and hundreds of smaller archipelagoes in the central Pacific Ocean, terminating more than 9,000 miles farther west at almost 140°E on a line of longitude that bites into the coast of South Australia beyond Tasmania. Farther north, its waters thread around Indonesia and the Philippines and slide onward past Japan. In 2005 it was estimated that twenty thousand demersal longliners worked WCPFC waters, setting lines for halibut, cod, and sablefish up into the Gulf of Alaska while their pelagic or midwater counterparts harvested tuna farther south.[11]

The WCPFC is the first of its kind in the western and central Pacific Ocean, and nations such as Australia, Canada, China, Japan, New Zealand, the United States—which joined in 2007—and the Pacific Island States of the Forum Fisheries Agency are implementing a comprehensive management program to conserve highly migratory tuna stocks. The commission has adopted measures to improve compliance with and enforcement of fisheries regulations under the UN Law of the Sea and its enabling agreements. That the WCPFC found it necessary to officially condemn illegal fishing suggests that it is dealing with operations by its own members. But each step forward matters. Importantly for its ability to take timely initiatives, the commission does not depend on consensus for its actions. A three-quarters majority of voting members can accept proposals and implement regulations. Seabird conservation specialists were heartened by the commission's 2005 Resolution on the Incidental Catch of Seabirds, which urged members to implement the FAO International Plan of Action—Seabirds and to furnish information about bycatch to the WCPFC Scientific Committee. Although resolutions are nonbinding, the 2005 resolution was also a step forward because all three North Pacific albatrosses spend a lot of time in WCPFC waters above 20°N and an additional fifteen more species glide and soar mainly below 30°S. All these birds together with their cousin shearwaters are susceptible to the tens of millions of hooks set by WCPFC longline fishers each year.

The following year, in Apia, Samoa, a WCPFC Voluntary Small Working Group on Seabird Bycatch Mitigation met to compare and evaluate the mitigation methods and devices other fishery management organizations were deploying. Australia, backed by New Zealand and following the CCAMLR blueprint, called for the mandatory use of thawed bait, a minimum length for streamer lines, weighted branch lines, and no offal or bycatch discharge during line setting, but members did not agree. They concluded that their bycatch measures needed to be flexible enough for commission members and vessel crews to test and then adopt measures and devices best suited for their fishing styles and locations.

New Caledonia–based fisheries expert Brett Molony reported that onboard observers had anecdotally recorded almost four thousand seabirds caught over the previous twenty-five-year period by tuna longliners in WCPFC waters. Most of the incidental deaths occurred in areas well known to bird specialists, and Molony was reporting a problem that WCPFC participants had known about from the start. They responded at their third regular session in 2006 by resolving to reconsider mitigation requirements based on application and testing by CCAMLR and "adopt measures for their own fisheries."[12] Mitigation measures are now required for longline fishers in the latitudes where most albatrosses fly, but how they are deployed and monitored remains an issue. The commission also adopted plans to manage fish stocks, head off illegal fishers, maintain a list of vessels known to fish illegally, and run a catch-documentation program.[13] The basic weaknesses BirdLife International sees in the WCPFC's accomplishments to date in regard to seabird bycatch are the gaps in data on seabird losses despite a measure that requires member states to submit such information annually to the commission and the absence of a fisher education program and a system for feedback from fishers. To address this need, bird experts presented papers to WCPFC's 2010 Scientific Committee meeting in Tonga reprising seabird bycatch issues and focusing on streamer effectiveness and improving sinking rates for pelagic longlines. An encouraging sign was the participation of Warren Papwith, the secretary of the Agreement for the Conservation of Albatrosses and Petrels (ACAP; see Chapter 22), which had signed a Memorandum of Understanding with WCPFC in 2007. He explored mitigation options and an onboard observer program based on a model developed by the Indian Ocean Tuna Commission and reiterated the ACAP objective of generating good data about fishery interactions with seabirds so as to promote specific mitigation measures. This kind of interaction among fishery bodies and ACAP is most promising.[14]

In 2003, the nine-year-old Commission for the Conservation of Southern Bluefin Tuna (CCSBT) expanded its membership to include a new category of "cooperating non-members," who act in liaison with the management body. These nonmembers have no voting rights but are willing and expected to abide by the objectives and regulations of the commission. Korea joined CCSBT in 2001, followed by Taiwan in 2002. Indonesia joined in 2003 under the newly defined nonmember status, and the Philippines, South Africa, and the European Union are also recent inductees to cooperative nonmember status.

CCSBT, based in Canberra, has an area of operations that includes all the major tuna fisher nations, and in 2007, about seventeen hundred vessels under seven flags appeared on the list of fishing boats licensed to take Bluefin Tuna in the CCSBT area. It also maintains a list of authorized farms for tuna and almost since its beginning has expressed interest in what happens to other marine animals as a result of fishing for tuna. The commission faces an enormous challenge in its goal of making Southern Bluefin Tuna fishing sustainable. CCSBT reduced the total allowable catch for 2007–2009 to 11,800 tons, with Australia permitted 5,300, Japan 3,000, New Zealand 420, and Korea and Taiwan 1,100 tons each. Nonmembers share fewer than 1,000 tons. Recently organized catch-documentation and vessel-monitoring schemes similar to CCAMLR's aim to curb illegal fishing in CCSBT's area. Currently, CCSBT is addressing the problem of illegal fishing and of vessels fishing for Bluefin Tuna under flags of convenience, and it requires members to deny entry to any bluefin shipment unless a recognized official of the export nation signs off on the vessel's name, gear, and catch area.

CCSBT suffers from significant weaknesses. Members often do not agree on catch limits or catch data, now admittedly underreported by Japan for many years. From the outset, members' fierce and loud arguments over catch limits created a rift that relegated seabird bycatch to minimal significance. All of Brothers's research and that of colleagues in other fisheries proved that the use of a mix of devices was significantly more effective in lowering seabird deaths than using one alone. But since 1995, CCSBT has only mandated the use of bird-scaring lines for vessels below 30°S, and it gathers little data to evaluate the effects of this requirement. Pressures are mounting among the membership to gather and analyze data not only for trade statistics but also for bycatch information. Cleo Small, BirdLife International's liaison with fishery management bodies, reports that disagreements remain about the need for onboard observers to monitor bycatch, and observers have been positioned on less than 10 percent of vessels operating under CCSBT regulations. Furthermore, observers do not have to report to the CCSBT, but to the vessel's national member, which may or may not share details with other parties. Debate also continues about how best to reduce illegal fishing.[15] Some CCSBT members like New Zealand are pushing the CCSBT toward expanded use of mitigation devices. Australia and ACAP have both argued that improved data would give the commission's Related Species Working Group the credibility to be taken seriously. It would allow a CCSBT consensus on sustainable harvests within ecological parameters. Now that fishing quotas have been lowered and harvests made more transparent, some members argue that bycatch has decreased, but the working group, which meets only every two years, is instructed to keep confidential the data it is gathering. In 2003, the group printed a ten-page pamphlet titled *Building a Seabird-Friendly Southern Bluefin Tuna Fishery*. Translated into five languages, it stresses the financial losses from catching birds instead of tuna, displays a range of mitigation devices, and provides instructions on how to use them, release birds alive, identify species, and report bycatch data. The Secretariat has worked with members intersessionally to improve the provision of observer information, and in 2007 the Related Species Working Group focused on whether it could develop a binding resolution to require members' vessels to report reliable bycatch data.[16] CCSBT has stated its intention to follow the FAO International Plan of Action—Seabirds and has mandated the use of streamer lines on vessels. It can build on these steps by requiring parties to supply data adequate to evaluating and improving bycatch mitigation. It is a good sign that other fishery management bodies and international bird groups have accepted invitations to participate in CCSBT's Ecologically Related Species Working Group.

Areas supervised by the twenty-seven-member Indian Ocean Tuna Commission (IOTC) and the International Commission for the Conservation of Atlantic Tuna (ICCAT) also rank high in respect to their significance to albatrosses. Three species, including all critically endangered Amsterdam Albatrosses, pass a high percentage of their breeding seasons in the IOTC area, and at least ten species breed in ICCAT waters. One of them is the mouse-harried endangered Tristan Albatross on Gough Island.

The IOTC's Working Party on Ecosystems and Bycatch concluded at its 2007 meeting that IOTC needed to do much more in its albatross-rich zone. Three years earlier it had passed a resolution calling for specific mechanisms to reduce seabird losses in the Indian Ocean to zero. But, as with CCSBT, data about the seabird problem are inconsistent and incomplete. Most members furnish no data to the IOTC Secretariat. It was clear to Australia that for any progress to be made, quantifiable data had to be collected regarding the effects of fishing on nontarget species. At the twelfth meeting in Oman in June 2008, the commission members mandated the use of at least two types of mitigation devices on tuna and swordfish longliners below 30°S latitude, as does the Western and Central Pacific Fisheries Commission. This pleased Australia, which had pushed for strengthening the bycatch mitigation requirements, and BirdLife called it a "highly positive step."[17] Compliance, however, is another matter.

The International Commission for the Conservation of Atlantic Tuna (ICCAT), the oldest of the fishery bodies with a significant albatross bycatch problem, provides the usual functions related to harvest in its area, which covers the Mediterranean Sea and the Atlantic Ocean. Based in Madrid, the commission currently has forty-eight members. It declares that unless another fishery body offers data, it will compile "data for other fish species that are caught in tuna fishing."[18] By this the commission means sharks. It makes no such declaration about seabirds. Although bycatch data are not provided regularly to the Secretariat, the commission has formed a subcommittee on ecosystems that examines bycatch issues in response to a request for a risk assessment, but risk assessment has yet to be completed. In a 2007 meeting, participants discussed bycatch data from fisheries off Brazil and southwest Africa, where about three thousand birds are dying annually from fishing activity in the Benguela Current. The group agreed to identify species at risk in convention waters, collect data about their distribution, and assess annual losses in terms of population trends over the longer term. The subcommittee also recommended that educational material be furnished to fishers, that the commission hire a bycatch coordinator and that its scientific delegation be enhanced with experts in seabird biology. A 2009 Seafood Sustainability White Paper noted that the bycatch group was encouraging members to implement national plans or at least implement elements within them, such as those elements adopted by Brazil, which has been gathering hard data on bird losses and also deploying mitigation devices in its fisheries. It notes that ICCAT requires longliners to fly streamers below 20°S during daylight, and commends Brazil for reducing bird losses. Clearly, there is a long way to go.[19]

Neither the Indian Ocean nor the Atlantic Tuna fishery management bodies has a strong record of gathering bycatch data or of practicing seabird bycatch mitigation, and neither has effective onboard observer programs. It has been suggested that nongovernmental organizations such as the World Fund for Nature and official bodies associated with BirdLife International and the Agreement for the Conservation of Albatrosses and Petrels offer expertise to these two fishery management bodies and supply information and training in bycatch mitigation. ACAP has proceeded to sign memorandums of understanding that facilitate exchanges between commercial and wildlife interests. For effective bycatch reduction to occur, the fishery bodies need to commit formally to monitoring bycatch and to educating vessel owners and operators in the most effective mitigation devices and procedures. Making these improvements is a slow process.

BirdLife International's Cleo Small has played a crucial role in working with regional fishery management bodies to develop seabird bycatch mitigation measures and regulations. She led a research team that compiled and edited a book about regional fishery bodies and albatross losses, *Regional Fisheries Management Organisations: Their Duties and Performance in Reducing Incidental Mortality of Albatrosses and Other Species*. She also participated in preparing the magnificent 2004 *Tracking Ocean Wanderers*, using the transponder-based research into albatross foraging and flight patterns of scientists like Henri Weimerskirch and colleagues to track and quantify the occurrences of albatrosses in the Southern

Hemisphere. Recent work has featured similar tracking of North Pacific albatrosses. This research, along with the data then available from global positioning systems, allowed Small to create overlays of fisheries and fishing activity with albatross presence, an essential first in understanding and managing the interactions between albatrosses and commercial fishing. Small has effectively introduced a collaborative role for BirdLife International into every regional fishery management body whose jurisdiction includes albatross nesting and foraging zones. Along with colleagues Ben Sullivan, Samantha Petersen, and Deon Nel, she has been well received as a partner in moving bycatch mitigation forward as a solid component of fishery management. Observers say that her success, like that of John Croxall, is the product of her skills in working strategically from the top down, assisting fishery management leaders in accepting and implementing the best measures today while being flexible enough to upgrade when more is known. Because BirdLife International and the Royal Society for the Protection of Birds back her, she has access to the best science available and uses this to move things along. Moreover, the crucial advocacy and action work of Greenpeace, the International Fund for Animal Welfare, and the Sea Shepherd Conservation Society create extreme positions along the conservation continuum that allow people like Small and fishery managers to achieve compromises among the membership. Small reports that after 2004, all five of the tuna management bodies started taking more notice of seabird bycatch, passing resolutions expressing concern and agreeing to consider measures to reduce bird deaths. Two of them began monitoring and evaluating the bycatch problem themselves.

Cleo Small is one of a handful of very skillful women committed to albatross bycatch mitigation. She has demonstrated wonderful abilities to work in highly collaborative ways, as has the United States' Kim Rivera, who is National Seabird Coordinator for NOAA Fisheries Service, and works closely with CCAMLR's Working Group on Incidental Mortality Associated with Fishing. Other colleagues such as Nicole LeBoeuf, Samantha Petersen, Pamela Toschik, Janice Molloy, and Rosemary Gales communicate effectively with the entire continuum of stakeholders, and set out to marshal the best science to support their work with fishers and fishery management bodies.

The collaboration between BirdLife International, with its global scientific and organizational resources, and the fishery management bodies is a huge step forward in marine ecosystem management. With the added component of ever stronger and more active support of Southern Ocean nations, fishery management bodies may be moving into a position of greater competence and effectiveness in protecting fish stocks, reducing bycatch, curbing illegal fishing, and maintaining the viability of marine ecosystems.

In fact, there is good news in that respect. Boris Worm of Dalhousie University and Ray Hilborn of the University of Washington, along with an international team of nineteen other authors, published in 2009 the findings of a follow-up study to the 2006 paper by Worm et al. in *Science* that analyzed a global trend toward fisheries collapse by mid-century. They found that in several fisheries intensively managed as ecosystems the rate of fishing had been reduced, resulting in some stock recovery. Their findings support the case for sound fishery management but also foreground the discrepancy in resources between wealthy and developing nations—for the most part, the fisheries showing some stock recovery were managed by wealthy nations. The exceptions included Kenya, where stock recovery had resulted from seasonal closures of fisheries and gear restrictions. Overall, the analysis suggested that fishing below what fishery management bodies have determined as "maximum sustainable yield" (MSY) yields as much fish as fishing beyond MSY but has important ecological benefits. In other words, fishing below MSY yields economic and ecosystem benefits at the same time. But the authors conclude with a caution: overfishing and stock depletion are still taking place in a large proportion of fisheries worldwide.

CHAPTER 18

Governments

BECAUSE IT IS NATIONS THAT ENFORCE multinational and international accords, it was an important step after World War II for albatross range states to claim and take possession of their Southern Ocean islands, rich in ocean-dwelling seabirds, endemic plants and insects, as well as birds, and of their coastal waters. In the mid-twentieth century, Australia, Chile, New Zealand, South Africa, France, and the United Kingdom began studying their Southern Hemisphere mid- to high-latitude terrestrial biomes and marine ecosystems and protecting and managing their specialized flora and fauna. In the North Pacific Ocean, the United States, Japan, and latterly Mexico began to declare some of their mid-latitude islands as wildlife refuges and to support research into the status and trends of terrestrial and marine wildlife. The environmentally responsible exercise of state sovereignty has been a powerful tool for saving albatrosses. It continues to play the essential role in ensuring progress in reversing the declines so evident in the oceans' ecosystems worldwide.

The political frame within which nations drew up and then implemented management plans for coastal and marine resources required that they reconcile the interests of fishers and the fishing industry's political clout with the conservative objectives of wildlife officials and agencies and nongovernmental environmental organizations. It also required responding to an informed public increasingly alarmed about a history of mismanagement of the environment and fragmentation of native plant and animal communities. Several albatross range states among the more than forty worldwide that harbor albatrosses have taken bold initiatives in international conservation organizations, both governmental and nongovernmental, and these states have also been leaders in developing and implementing national plans of action to reduce bycatch and conserve seabirds.

Australians pride themselves on their progressive environmental citizenship and leadership. By the mid-1990s, Australia was a signatory to sixty-six international conservation treaties, conventions, and protocols, half of which related to the marine environment. Australia has a strong interest in albatross conservation—for good reason. In addition to about twelve thousand pairs of its endemic Shy Albatross that nest on three islands in Tasmania (two of which lie in the Tasmanian Wilderness World Heritage Area), four additional species breed on its two World Heritage sub-Antarctic islands HIMI and Macquarie. Furthermore, all Southern Hemisphere species, with the exception of Ecuador's Waved Albatross, reportedly feed within Australian national waters. Even the Laysan Albatross from the Northern Hemisphere has been recorded on Australia's Norfolk Island.

Australia's interest and leadership in regard to Antarctica began with Douglas Mawson's 1911 expedition and led in 1947 to the formation of the Australian Antarctic Division. This division operates in accord with the purposes of the Antarctic Treaty System. It is a complicated policy agency with competing objectives and a confusing administrative structure. It represents Australia at Antarctic Treaty System meetings and other conferences that deal with South Pole issues. It administers Australia's Antarctic territories and sub-Antarctic islands and conducts research on the Southern Ocean, maintaining three permanent stations in Antarctica. Antarctic Division personnel monitor albatross and other seabird colonies, track migratory movements, study global environmental changes, and analyze the effects of fishing on seabird communities in sub-Antarctic waters.[1]

As an original signatory to the Antarctic Treaty and a party to supplemental agreements, Australia grounds the protection of its sovereignty and national security in a regional context. Accordingly, Australia has been outspoken in political debates regarding international environmental regimes in the Southern Hemisphere. In 1989, with strong nongovernmental backing, Australia led a revolt with France and New Zealand against those who wanted to plan future mineral development in Antarctica through a mineral resource activity convention that opened for signatures in Wellington. New Zealand's push for a prohibition of such activities and Australia's refusal to sign the new treaty led to a fifty-year moratorium on all mining on the Antarctic continent. There is concern over the kinds of mineral exploitation that may occur once the mining moratorium expires, and Australians are wary of the new scientific stations that China, India, South Korea, and others have established on Antarctica, for a total of more than fifty bases.[2]

Australia had been managing its 200-mile fishing zone for twenty years when the UN Law of the Sea entered into force in 1994. Encompassing 3.4 million square miles along a coastline 23,000 miles long, it is the world's largest marine area claimed by a single nation. Australia has also declared exclusive economic zones around HIMI, Macquarie Island, and other offshore islands as well as off its claims on the Antarctic continent. As a federation with delineations of power between the central government and states and territories, the Canberra-based government deals with external affairs, common standards, and funding for conservation-related programs. Each state has authority over its immediate waters, regulating fishing in the first 3 nautical miles, at which point the commonwealth government takes control out to the 200-nautical-mile limit. The commonwealth government also manages fisheries that cross more than one state's waters. In consultation with the Antarctic Division, which manages the islands, and with fishing representatives, the Australian Fisheries Management Authority conducts a five-year rolling research program in the HIMI fishery. In making fishery and seabird management decisions, Australia has for the most part aligned itself with the intent and provisions of the multinational regimes that it has joined, such as the Convention on the Conservation of Antarctic Marine Living Resources and the Commission for the Conservation of Southern Bluefin Tuna. As a result, Australia offers these fishery bodies an effective national enforcement arm they would otherwise lack.

Australia designated Macquarie Island a Nature Reserve in 1978, it became a World Heritage Area in 1997, and in 1999 Tasmania added a marine extension. The Australian government also placed this new marine reserve into Australia's National Representative System of Marine Protected Areas under its 1998 Oceans Policy. Zoned for biodiversity, habitat protection, and long-term ecological viability, the marine reserve encompasses about one-third, or 6,200 square miles, of the economic zone around Macquarie. The marine sanctuary is one of Australia's dozen Biosphere Reserves inscribed under UNESCO's Man and the Biosphere program. Scientists worry about how human activities may affect the approximately 3.5 million seabirds and one hundred thousand seals that rely on Macquarie's water and land for food and reproduction. There are thirty-eight bird species on Macquarie, including nineteen threatened seabirds, such as Wandering, Grey-headed, Light-mantled, and Black-browed Albatrosses. Fishing is not permitted inside the reserve. A demersal trawl fishery for Patagonian

Toothfish operated in the economic zone to the west of the island under a regulated permit from 1994 through 1997, but a rapid decline in stocks led to a reduced catch limit and careful monitoring of further fishing. The Antarctic Division continues to be concerned about damage from trawlers to fish stocks and the seabed around Macquarie Island. Of pressing concern on land, the removal of invasive species has become a vexed issue on Macquarie. In May 2010, a final costly and prolonged effort to eradicate predatory rats and introduced house mice and rabbits that have degraded vegetation and caused soil erosion began with one ship from Hobart with a crew of twenty, four helicopters, and three hundred tons of bait bearing southeast for the 850–mile run to Macquarie. This campaign followed an earlier one successfully waged against feral cats that preyed on seabirds. They still do. ACAP data show that cats threaten eight albatrosses in fourteen sites, compared with rats that prey on three albatross species. However, on Macquarie, both rats and notably rabbits, freed from feral cats, have increased over the past several years. With important NGO and ecotour support, Australians began working to return the heritage site to something resembling its pristine state. But the baiting team in 2010, challenged by fog and poor weather that also grounded the helicopters, was able to deliver only 8 percent of the poison. Although David O'Byrne, minister for Environment, Parks and Heritage, remains committed to seeing the island restored to an "ecosystem free of introduced pests,"[3] he recalled the team to Hobart and put off the A$24.6 million jointly funded project for another year.

In 2002, Australia declared the north and northeast sea quadrants of its economic zone around HIMI as one huge marine reserve of twenty-five thousand square miles, placing waters off-limits to all fishing. The reserve is the world's second largest wholly protected marine sanctuary, exceeded only by Australia's Great Barrier Reef Marine Park. By agreement, the entire HIMI reserve is also subject to CCAMLR's regulations, which are greatly strengthened by Australia's track record in enforcing them.

Since the mid-1990s, illegal fishers have taken advantage of the remote and uninhabited character of Heard and McDonald Islands to plunder native toothfish. Illegal entry and harvests by demersal toothfish operators have been costly to the health of the fishery. In spite of the high cost of surveillance of waters that are thousands of miles from the Australian mainland, Australian fisheries vessels have intercepted and detained foreign longliners for taking toothfish illegally in those waters. In 2003, Australia arrested a total of one hundred vessels for fishing illegally, mostly in its northern waters. Australia and New Zealand were at the time using an armed civilian vessel in a program called Operation

Rushwater to intercept illegal toothfish operations. One incident, that of the Uruguayan *Viarsa 1*, became an international chase.

If you want to understand the shell game of illegal fishing, consider the history of the *Viarsa 1*. Built in Ise, Japan, she bore two different names prior to taking *Viarsa 1* from her Panamanian company. Then a Spanish owner in La Coruña leased the *Viarsa 1* to a Uruguayan firm that sent the vessel into the Southern Hemisphere for a spell of fishing. From 2000 to 2003, she is recorded to have offloaded Patagonian Toothfish in Port Louis, Mauritius, a major pirate port for illegal toothfishers.

On August 7, 2003, Australian authorities sighted the *Viarsa 1* allegedly fishing without a license off HIMI. The customs vessel *Southern Supporter* challenged the twenty-three-year-old fishing boat, but instead of heaving to, the *Viarsa 1* sped west and sparked the longest chase in Australian maritime history. The *Dorada*, a British fisheries vessel out of the Falklands, and a deep-sea tug from South Africa joined the chase. Dodging icebergs, enduring pounding seas, and crashing through sea ice, the longline vessel, dogged by its three pursuers, steamed on. After a twenty-one-day chase, the *Viarsa 1* eventually slowed, and Australian and armed South African personnel boarded her, arrested her forty-member crew, and impounded what turned out to be 150 tons of Patagonian Toothfish worth A$2 million. On October 3, the *Southern Supporter* accompanied the *Viarsa 1* into Fremantle, the nearest Australian port, and authorities seized the longliner.[4]

In Montevideo, company officers claimed their vessel was fishing off South Africa when sighted. However, they paid a fine for the vessel's having a disabled CCAMLR-required monitoring system. Four months later, Uruguay pulled the license of the Uruguayan leasing company, and the boat remained in Fremantle with five of her crew. Charged under the 1991 Australian Fisheries Management Act, the Uruguayan skipper, one Chilean, and three Spanish crewmembers stood trial in Western Australia's district court. To the dismay of Australian officials, the nine-week case ended with a hung jury. In a retrial in September 2005, the Perth district court cleared the crew of fishing illegally on the grounds that Australian interceptors had not actually observed the vessel fishing in national waters. By then the crew had been detained in Australia for more than two years. The *Viarsa 1*'s owners filed a counterclaim, which the court denied them, for A$10 million for damages and lost revenues.

But the court's decision to clear the crew of illegal fishing puzzled and embarrassed maritime lawyers. Putting on a bold face, Fisheries minister Ian Macdonald issued a press release to emphasize Australia's commitment to stop illegal fishing by reporting that the government had acquired a permanently armed

Southern Ocean patrol vessel, the *Oceanic Viking*. He concluded that the Howard government was absolutely committed to defending Australia's fish stocks and marine environment and would take every possible action to defeat organized criminal activity in its fishing zone.[5]

Both Australia and Britain have referred to the *Viarsa 1* in CCAMLR discussions as evidence that the UN Law of the Sea needs the support of binding regulations and enforcement by fishery management organizations. They began publishing a blacklist of vessels that take fish illegally, and the Coalition of Legal Toothfish Operators (COLTO) joined them in this effort.[6] The case of the *Viarsa 1* is by no means unusual. In the 2002–2003 Patagonian Toothfish fishing season, at least eight thousand tons reportedly were poached in CCAMLR's zone, and another ten thousand tons in illegal catches likely came from the same waters. Toothfish sales topping US$100 million in the United States alone every year provide strong incentives to get the fish to market, no matter the risk or cost.

In June 2007, through a Fisheries Legislation Bill, Australia's Lower House strengthened oversight of offshore waters, for both national security and to protect fisheries. It committed A$800 million to the effort, and the legislation added the penalty of vessel forfeiture to a list of improved policing and information-sharing measures. As a result, Australian authorities have been able to create a number of human-made reefs by sinking confiscated vessels. Fisheries Secretary Susan Ley reports that fishers seemed to have got the message.[7]

Australia joined the Convention on the Conservation of Migratory Species of Wild Animals in 1991, and in its first formal report to the parties that year alerted the convention to the seriousness of seabird deaths in longlining. It also commissioned Rosemary Gales in 1991 to conduct a review of all albatross species. Gales's 1993 report was sobering, followed by equally alarming reports from the seabird scientists at the first World Albatross Conference in 1995. In that same year, Australian officials declared four of Australia's five nesting albatross species as Vulnerable and the fifth, the Wandering Albatross on Macquarie Island, as Endangered. Their populations were declining not only from longlining at sea but also from predatory animals and habitat degradation on their nest islands. At the 1997 CMS meeting of parties, Australia nominated the Amsterdam Albatross to the convention's Appendix I, to join the Short-tailed Albatross listed in 1979. This appendix lists migratory animals in danger of extinction and requires range states not only to prohibit their capture but also to grant the species full protection. At the same time, Australia named an additional eleven Southern Hemisphere albatross species to be listed in Appendix II, which lists species with an unfavorable conservation status and encourages range states to cooperate through regional conservation agreements. Biologists and government officials hoped that listing nine species of the Great Albatross and two species of the Sooty Albatross would alarm members into taking action. Australia repeated Gales's conclusion that the future of the albatross family depended on tailoring international cooperative agreements for their protection. Australian Environment minister Robert Hill believed, correctly, that the CMS listing was "an important starting point" in global cooperation for albatrosses that would "provide a mechanism for southern hemisphere countries, linked by the great southern oceans, to work together to conserve the species."[8] He was right. From that time forward, bycatch was on the agenda at CMS meetings, eventually leading to the formation of the 2001 Agreement for the Conservation of Albatrosses and Petrels. Greenpeace congratulated the Australian delegation for its additions to the CMS appendices but criticized New Zealand, not a member of the convention, for failing to be proactive about albatrosses and other seabirds despite being host to so many nesting species. New Zealand eventually joined the CMS in 2000.[9]

When Australia listed the endangered albatrosses on the CMS list in 1997, it also called for the development of a national threat abatement plan for seabirds. The abatement plan team relied on input from seabird scientists, personnel from the Australian Fisheries Management Authority, domestic tuna longline representatives, and the Humane Society International. The final plan called for a reduction in seabird bycatch, to 0.05 birds per 1,000 hooks, an overall 90 percent decrease over a five-year period at a cost of close to A$1 million. It recommended modifications for fishing gear and practices and training for fishers.

At the time, Brothers, Gales, and Robertson were part of a collaboration focused on albatross and petrel recovery chaired by Barry Baker, then with Environment Australia's Natural Heritage Division. Recognizing the groundwork the abatement plan had accomplished, they built on it, adding a component for the management of breeding islands to eliminate pests and restore habitat. Their plan pointed to trawling as a serious threat and called for the collection of reliable data from trawling. It called for virtually the same mitigation devices and procedures in Australian waters as CCAMLR required, as well as education for fishers. It included scientific monitoring of thirteen populations of birds considered nationally Vulnerable and four species considered Endangered. Even though Australian hooks represented only about 10 percent of all hooks set around the globe each year, Australian domestic longlining vessels were setting roughly 100 million hooks per year, and the group believed their crews could significantly reduce the number of bird deaths. The plan called for the same lowered bycatch rate as the abatement plan but with a price tag seven times higher. The final *Recovery*

Plan for Albatrosses and Petrels listed twenty-one recovery actions at a cost of A$7.5 million over a five-year period. It was released for public comment in November 1999 and the minister of Environment signed it in October 2001.[10]

The *Recovery Plan* has served as a blueprint for other range states with breeding and foraging species as they draw up their National Plans of Action for Seabirds. In addition to implementing and monitoring this plan, Australian biologists continue to engage in basic research and important fieldwork on Australia's island breeding sites. Most of the nest sites are now protected by both national and international regulations, and access is restricted. Analysis of efforts to eradicate introduced predators on albatross islands revealed some interesting cost comparisons. Flesh-footed Shearwaters were being lost in the tuna and billfish fishery off Australia's east coast around Lord Howe Island, the bird's sole breeding site. Calculations showed that closing the fishery would significantly reduce bird deaths but at a cost of A$3.5 million in fish not caught. By comparison, an all-out effort to rid the island of rats, predators of shearwaters, would cost slightly more than A$.5 million and result in a 30 percent increase in the Shearwater colony population.[11]

New Zealand claims the fifth-largest territorial sea in the world, blanketing just over 1.5 million square miles. It is fifteen times larger than the mainland, which is about the same size as the United Kingdom or the US state of Colorado. Located in a zone of convergence, the seven sub-Antarctic islands on the underwater New Zealand Plateau provide essential habitat for seabirds. Almost one quarter of the world's seabird species, totaling perhaps 25 million individuals, breed in New Zealand, including thirteen albatrosses. Thirty-five seabird species, including seven endangered albatrosses, nest exclusively in New Zealand. Deservedly then, New Zealand takes pride in being the seabird and albatross capital of the world and has laws in place to protect albatrosses and most other seabirds. One would think that having such a natural heritage and laws in place to protect it would engender unqualified admiration and respect. However, there is a complicating factor. New Zealand also operates a world-class commercial fishery. At the turn of this century, fishing vessels landed more than 550,000 tons of saltwater fish annually, valued at about US$1.3 billion. Fish are New Zealand's fourth-highest-earning commodity, just after farm produce, and exports generate about one-half of the fishing income. In laying six thousand miles of nets, setting fifty million hooks, and operating ninety thousand trawls, commercial fishing employs about twenty-six thousand workers who catch or process some forty fish species, the key ones being deepwater Orange Roughy and midwater hoki, ling, hake, and squid.[12]

After declaring an exclusive fishing zone in 1965, the New Zealand government offered guarantees and subsidies for fishing vessels and equipment, sparking the growth of its modern fleet of fifteen hundred vessels. Further expansion after 1978, when foreign vessels shifted to deeper water, led to joint ventures with the USSR, Japan, and other Asian countries. New onshore processing plants led to the doubling of fish exports during the 1980s. By the end of 2010, the industry expects to generate a steady NZ$2 billion from annual export revenues.[13]

A sensitive but questionable trade-off, according to some scientists and officials, takes place between the Ministry of Fisheries' adherence to sustainable use through catch quotas and the Department of Conservation's commitment to wild animal management. One of the costs of this trade-off is the resistance of the fishing industry and agency to sharing data with conservation agencies and organizations. This failure to communicate causes friction. The Royal Forest and Bird Protection Society, an advocacy group with forty thousand members that partners with BirdLife International, has criticized the government's uneven and limited record in ocean conservation, condemning its marine policies as one-sided. For instance, in 2007, the group denounced the government's NZ$2 million break on fishing levies in return for trawlers' "agreeing not to destroy the seabed in areas where the industry never had any intention of bottom trawling."[14] It accused the seafood industry of "greenwashing," by suggesting that the area under a bottom-trawl ban would be a superlative marine reserve. "Most people would assume that there would be no fishing allowed in a marine conservation area, but this isn't the case," bird representatives grumbled. "The fishing industry will still be allowed to fish in these areas, which can hardly be described as marine conservation."[15] Forest & Bird also challenged a requirement that New Zealand–flagged vessels adopt mitigation devices when fishing in distant seas but not in their own waters, when as many as ten thousand albatrosses and petrels have been dying annually in home longline fisheries.[16] Forest & Bird has worked to resolve such contradictions, but it says it encounters obstacles in the government's protective stance toward fishing interests. Effective long-term protection of the fishing industry has been shown to include generating and sharing data on fish and bycatch and the adoption of mitigation methods most suited to a specific fishery. In contrast, New Zealand's protection of the fishing industry has meant separation and contradiction, not integration of government objectives and efforts. This has severely limited the effectiveness of laws and efforts to protect seabirds and albatrosses, despite the unique wealth of New Zealand's breeding seabird populations. At the June 2010 World Oceans Day, advocates recognized how valuable New Zealand's ocean ecosystem is in the regional and global community and suggested it was time to give the natural treasure of native sea life its due by pressing for improvement in the fact that "just 0.3% of New Zealand's marine environment is protected in marine reserves, compared with more than 30% of our land area that is protected." A consortium of agencies and organizations has set out to increase the area of protected marine environment.[17]

Albatross conservation figures into this natural wealth not in efforts to save populations through bycatch mitigation but in efforts to restore island habitat, still severely damaged and threatened by introduced species of plants and animals. For over a century, colonial settlers had worked to acclimatize nonnative plants and animals in an effort to "improve" the native flora and fauna of New Zealand and Australia. But awareness of the threats presented by some of these alien species surfaced in the 1950s, especially the threat of rats and mice to albatrosses and other seabirds. The Pacific Rat had landed on some of the remote Southern Ocean islands from visiting ancestral Maori boats. A voyager records seeing rats scurrying ashore from James Cook's exploration ship the *Resolution* when it was beached for repairs in New Zealand's Fiordland in 1773. Ships' rats and mice gained access and spread during the sealing and whaling era in the 1800s on New Zealand's main and offshore islands with seabird nest colonies. Today rats exist on a good 80 percent of the world's major islands, including at least one-third of the New Zealand archipelago, and experts deem them to be major factors in the destruction and extinction of both land and sea birds. Lance Richdale's work at Taiaroa Head was the earliest attempt to protect a nest site from such predators, and in the 1960s, the New Zealand government began sporadic efforts to restore near-shore breeding islands because of their role as genetic diversity banks. Forest & Bird laid out rat poisons from 1960 through 1964 on tiny Maria Island to protect storm-petrels. That and other efforts had some success in reducing rodent numbers, but at the time nobody believed much more could be done. In ensuing years, workers conducted eradication programs on other seabird nest islands, but in general the program was pretty hit and miss, with successes on small sites that could be hand sown using poison or traps.

In 1976, researchers and officials gathered for a symposium on rodent control in New Zealand Nature Reserves argued that although complete extermination was the only solution, it was a near impossibility. However, at about the same time a second generation of anticoagulant poisons became available that rodents consumed in lethal doses before showing symptoms. Officials began experimenting with bait stations and silos with the improved attractants and spelled out measures for avoiding killing nontarget species. These developments allowed personnel to move thinking forward from planning sporadic control toward an ideal of complete eradication of the most threatening species from the larger nesting islands. Today almost one hundred islands are clear of rodents. Experts have refined the procedures for maximum efficiency, dropping "one poison one hit" baits along preestablished grids, which has made it possible to tackle the larger islands. Eventually aerial spreading of the new anticoagulants became cost-effective, and by 1992, helicopters were part of the invasive-animal control arsenal.

In 1992, workers began to kill rabbits and mice on Enderby Island in the Auckland Island group, where an increasing number of Southern Royal Albatrosses nested. The Victoria Society of Australia had released rabbits on Enderby under a European species acclimatization program, and they were threatening the nesting Royals. Workers trapped fifty of the rabbits and shipped them off to North America as a rare breed, where they became fashionable pets. Personnel finally removed the few cattle that were shambling about and living mostly on kelp, and a year later, two airdrops using anticoagulant baits poisoned the rats and almost all the remaining rabbits. A trained dog, spotlighting, and traps removed the holdouts. Vegetation has returned and a record sixty-nine pairs of Southern Royals nested on Enderby in 2001. They are holding steady and possibly increasing.[18]

Campbell Island, designated a reserve in 1952, was the largest island targeted by an eradication program. The Southern Royal Albatross breeds only in New Zealand, and 99 percent of almost eight thousand breeding pairs have colonized Campbell Island. Overall, the population may be around fifty thousand birds, with some twenty thousand pairs breeding annually (see Table 1). At twenty-eight thousand acres, Campbell is the second largest of New Zealand's sub-Antarctic groups, one-fifth the size of the Auckland Islands. In the early 1890s, New Zealand officials leased Campbell Island as a sheep run. Scottish shepherds tended more than eight thousand woollybacks until most herders quit when the economic downturn in 1931 depressed wool prices. In doing so, they abandoned four thousand sheep and forty head of cattle, which continued to graze down the tussock grasses albatrosses use to build their nest pedestals. Rooting hogs and nimble goats from the settlement period had also worn out the native grasses. The alien Norway or Brown Rat had built up a population density reportedly unequaled for rats anywhere in the world.

Conservation officials eliminated the cattle by 1984, the feral sheep were finally removed from the island by 1992, and feral cats eventually died out. But in 1998, when Campbell Island and New Zealand's other sub-Antarctic islands received World Heritage Site status, rats were still plentiful on Campbell. In 2001, using helicopters, pest workers laid down one hundred twenty tons of bait in order to poison an estimated two hundred thousand rats. It worked. Although it took two seasons and cost almost NZ$3 million, by 2003 Campbell was rid of its rats. New Zealanders became world leaders in eradication programs, and Kiwi teams began to travel to train and assist workers in pest eradication on the French islands with some success; on the Falklands and South Georgia in the Southern Ocean, where rat poisoning will begin in 2011; as well as recently in Canada and the central South Pacific. Today, Campbell Island is a showcase for Kiwi expertise in controlling and eradicating alien animals and for restoring offshore habitat and reintroducing native species.

In 2000, the New Zealand Department of Conservation and its Seafood Industry Council, Ministry of Fisheries, and minister of Foreign Affairs and Trade, joined by the FAO, the IUCN, fishing gear manufacturers, and other interested parties, hosted the First International Fishers Forum (IFF1). Convened by the Department of Conservation's Janice Molloy, the express purpose of the four-day meeting was to develop a coordinated global response to seabird bycatch in commercial fishing. This included coordinating research, education, and training to avoid duplication and providing fishers with information on using seabird population modeling programs to predict the effects of specific fisheries on seabirds in their areas. Three others followed this forum; the fourth was in Costa Rica in 2007. In each instance, stakeholders from around the world have come together in differing combinations to sponsor, support, and participate in these forums.[19]

In August 2003, Fisheries minister Peter Hodgson released a draft National Plan of Action to Reduce the Incidental Catch of Seabirds in New Zealand Fisheries. Commercial longlining, trawling, and set netting for a dozen native fish species reportedly had killed thirteen species of albatrosses and seventeen species of petrels. Forest & Bird estimated that sixty-five thousand albatrosses and petrels had died on longlines set for ling and tuna in New Zealand waters over the twenty-year period after 1987, leading to a 35 percent decline in Salvin's Albatross and almost double that for Gibson's Albatross, accepted as a subspecies of the Antipodean Albatross. The group concluded that ten thousand albatrosses and petrels were lost each year in the nation's exclusive economic zone.[20]

The final 2004 National Plan of Action for Seabirds drew praise for its recommendations:

- A code of agreed best practices specific to each fishery to reduce bird bycatch
- The use of bird-scaring lines in longline tuna fisheries and the abolition of net sonde monitoring cables in trawl fisheries
- Education and training for fishers

- Sanctions and penalties for operators who failed to exercise reasonable care in reducing bird casualties
- Suspension of fishing in a specified fishery after vessels exceeded a seabird bycatch limit[21]

This is a strong program on paper, but the *National Plan* side-stepped the question of compliance by not spelling out whether it was voluntary or mandatory, which meant it was voluntary by default. Voluntary compliance was an important component of the government's relationship with its fishing industry, which argued that voluntary compliance allowed codes of best practice to remain flexible and lend scope for innovation. Moreover, fishers knew best how to protect seabirds in ways that were cost-effective. The plan's combination of self-regulation with only a threat of mandated controls pleased the fishers and drew fire from conservation agencies and organizations. Without mandatory requirements, they argued, the government had no authority to prosecute reckless, noncompliant, and illegal fishers. The Auckland-based nonprofit TerraNature Trust condemned the plan's voluntary approach and demanded stiff penalties for fishers who were clearly failing to comply with mitigation requirements. With the majority of the world's albatrosses in serious trouble, including New Zealand's endemic birds, the trust demanded that the code of best practices be not only mandatory but also imposed immediately.[22]

Despite these weaknesses, New Zealand's *National Plan of Action* established standards for her flagged vessels in protecting seabirds in home waters and on the high seas. Back-of-the-boat observation of actual, compared with estimated, seabird losses has been a vital component of New Zealand's plan. Officials calculate bycatch probabilities in specific fisheries and set loss thresholds. Counting and identifying carcasses down to the species level and the mode of capture, they compare actual figures with those of their models to improve the models and the mitigation methods.

A government review of marine oversight in New Zealand's 2003 *Oceans Policy* had acknowledged the obstacle of long-standing ideological difference between the Department of Conservation and the Ministry of Fisheries. Their conflicting objectives severely hampered sound marine management, slowing pollution cleanup and blocking the establishment of marine protected areas.[23] In 2005, a preliminary report for a new *Oceans Policy* suggested that the voluntary approach for domestic fishers be continued. It recommended that agencies charged with overseeing fisheries, ocean dumping, seabed drilling, and so forth fulfill their environmental obligations by conducting environmental assessments of the effects of their operations. But voluntary compliance and industry-prepared environmental assessments were not what Forest & Bird and other nongovernmental organizations wanted. However, an unexpected action in May of that year by the minister of Fisheries, David Benson-Pope, changed that voluntary position. He suddenly ordered most of the thirty-five-vessel squid fishing fleet into port to stop its killing of seabirds. Noting that "the squid fishing industry has had every opportunity to act responsibly and despite some good operators the majority have chosen not to," the minister explained that poor compliance with the code of best practice had forced his hand. He directed that observers be placed on most vessels, largely foreign-owned but chartered to New Zealand companies and operating in national waters, and mandated the use of mitigation devices. He threatened to impose substantial fines on any boat that failed to comply. "These measures," said Benson-Pope, "are the inevitable consequence of their poor behaviour."[24] This abrupt government move was positive and confirmed critics' fears that voluntary compliance had not stopped, nor was it likely to stop, bird losses.

In February 2008, Fisheries minister Jim Anderton finally mandated a set of what he called "simple measures"[25] proven to be effective in comparable overseas fisheries. They included the use of bird-scaring streamers by all trawlers and discharge of offal only at times of low opportunity for birds to be destroyed by the trawl cables. Longliners were to fly streamers and set lines at night or use weighted lines, and crews were to discharge offal away from baited hooks. Anderton acknowledged that these measures would impose costs on the industry and technical difficulties for some fishers. But taking no action and allowing ongoing injuries and deaths for significant numbers of seabirds was unacceptable.

Like Australia, New Zealand has also sent long-range aircraft to scout for illegal fishing in its sub-Antarctic sector. In mid-December 2005, a P3K Orion returned from a twelve-hour flight to the edge of the Antarctic sea ice and back to report a number of vessels to be checked on. New Zealand is a member of the High Seas Task Force and in collaboration with other members is working to exert leverage on illegal fishers. This means intercepting vessels on the high seas and publishing the names of proven miscreants. The United States has agreed to draw up a list of vessels fishing illegally both in and outside its territorial waters. If the resources to generate a comprehensive vessel and crew list can be marshaled, then such a document in the hands of the UN would link Northern Hemisphere efforts to halt illegal fishing with those in the Southern Hemisphere and support the development of a global surveillance system to identify, track, and apprehend illegal fishers.[26]

The four Northern Hemisphere nations with jurisdiction over major nesting places for albatrosses have also been forceful in efforts to reduce bycatch and stop illegal fishing. The UK's Overseas Territories of Tristan da Cunha and Gough host the endemic Tristan, Atlantic Yellow-nosed, and Sooty Albatrosses. The Falkland Islands are home to up to 80 percent of the world's nesting Black-brows, and South Georgia boasts major nesting colonies of Wandering and Grey-headed Albatrosses. Large numbers of seabirds breed or forage in the Southwest Atlantic sector of the Southern Ocean, including one-fourth of the sixty or so species reported killed on longlines.

Four species of albatrosses have been recorded as bycatch in Falklands waters that support more than sixty-three seabird species and twenty-nine nesting seabirds. The Black-browed Albatross, the sole nesting albatross on the Falklands, breeds at twelve sites on the archipelago. A sudden, startling loss of Black-brows in the first few years of this century from one census to the next drew an international response. Officials relisted the species from Near Threatened in 2000 to Endangered in 2003 (see Table 5). In 2004, under contract to the UK's Royal Society for the Protection of Birds, and with support from the Falkland Islands Fisheries Department and local people, Falklands Conservation drafted national plans of action for reducing bycatch, one for longlining and one for trawling. A first for a UK overseas territory, these plans reflect collaboration between local officials and officials in London. They address the problem on a scale that reflects local control while treating seabird protection as an international problem, involving neighbors from South America. Because of its own legal obligations to ensure that international law is respected in an Overseas Territory, the United Kingdom supports the plan. As a Northern Hemisphere nation, it exerts significant influence on behalf of the albatross through the regional treaties and conservation agreements in which it participates.[27]

The Falklands National Plan of Action for Seabirds in Longline Fisheries demonstrates close collaboration between officials, environmental groups, and commercial fishers. It includes extraterritorial links with South American nations to conserve Falklands breeding birds, especially the Black-browed Albatross. The plan aims to reduce longline mortality to 0.02 or fewer birds per 1,000 hooks. An observer program is in place, and when a vessel takes more than ten birds in a fourteen-day period, extra measures kick in. A suite of operational procedures and technical measures includes weighting lines and experimenting with faster sink rates, and experts agree that fishing between nautical twilight and dawn is the most effective way to reduce seabird losses in those fisheries. Restricting longline fishing on South Georgia to the non-nesting months from May

213

TABLE 5. Changes for Eight Albatross Species

SPECIES	STATUS IN 2000	STATUS IN 2010
Tristan Albatross	Endangered	Critically Endangered, with 80% pop. decline over 70 years due to longlines and invasive species on nest islands.
Waved Albatross	Vulnerable	Critically Endangered due to inshore and longline fishing, disturbance and invasive species on Galápagos.
Atlantic Yellow-nosed Albatross	Near Threatened (maybe qualifying for Threatened status)	Endangered (facing a very high risk of extinction), with rapid decline largely due to longlines (Brazil and southern Africa).
Black-browed Albatross	Near Threatened	Endangered since 2003, with 50% reduction over approx. 65 years: species most often drowned in longlines.
Black-footed Albatross	Vulnerable	Endangered since 2003, with projected rapid decline over approx. 65 years due to driftnets (banned), longlines, plastic ingestion.
Indian Yellow-nosed Albatross	Vulnerable	Endangered since 2003, with more than 50% reduction over approx. 71 years due to longlines (southwest Australia, southern Africa) and disease (Amsterdam Is., avian cholera)
Laysan Albatross	Least Concern (likely widespread and/or abundant)	Vulnerable since 2003, with probably 30% reduction over approx. 80 years; likely relisted as Near Threatened due to longlines.
Sooty Albatross	Vulnerable	Endangered since 2003, with more than 75% reduction over 90 years: may be relisted as Critically Endangered due to longlines and invasive nest island species.

Source: www.iucnredlist.org (2010).

Note: The status of thirteen additional species remains unchanged (unless now recognized as a separate species) in the decade. Only Buller's Albatross is reportedly numerically stable or slightly increasing. All other species are in various rates of decline.

to September has also been effective in reducing seabird deaths, as it has been around South Africa's Prince Edward Islands. Based on South Georgia data, the Falklands plan calls for adjusting the fishing season to the breeding cycle of the albatrosses and dividing Falkland waters into two areas, the Burdwood Zone and the Falkland Zone, to correspond to the seasonal abundance of breeding birds. Because tests in the trawl fishery in the 2002–2003 season counted more than 1,500 Black-brows struck and killed by the warp cable attached to the trawl, the plan mandates the use of streamer lines in trawling to keep birds off that particular cable. A new action plan for seabirds associated with trawl fisheries appeared in 2010 and is expected to build upon significant reductions in mortality and to keep a dialogue going between bird experts and fishery interests.[28]

New Zealander Derek Brown directed an eradication program for invasive animals, notably rats, on nine islands in the Falklands group. Both Brown and Black Rats inhabited places where about four million seabirds representing twenty-two species were breeding. In winter 2001, New Zealanders conducted a pilot project for four small tussock-grass islands close to Stanley, capital of the Falkland Islands, with the result that a year later, no signs of rats remained. Other sites have been baited, and results are promising, although care must be taken to avoid poisoning scavenging vultures and caracaras. A check on the largest poisoned island in 2006 turned up no rats and a greater diversity of native birds than identified before clearance three years earlier. Currently twenty-two islands have been cleared, but this is a long way to go in that splintered archipelago. At least seventy-seven more islands possess alien animals while hundreds of remaining islands and islets have yet to be surveyed.[29]

French biologists on Crozet, Amsterdam, and Kerguelen monitor the activities of seven breeding albatrosses on these islands and in the surrounding Indian Ocean. France informed members at the 2004 CCAMLR meeting of its intention of reducing seabird bycatch mortality around Crozet and Kerguelen. Bird deaths on the French islands had been alarmingly and embarrassingly higher than in other Southern Ocean zones. In fact, seabird bycatch rates had sunk to record lows across the entire CCAMLR region except for waters belonging to France. To some extent the French delay in responding to the threat from bycatch, habitat degradation, and alien species reflected the French national emphasis on theoretical at the expense of applied or practical science. With pressure and support from other CCAMLR colleagues, the French eventually were able to reduce seabird losses around Kerguelen and Crozet by 75 percent. In 2007, a nongovernmental alliance of more than one hundred environmental organizations with interests in the region, the Antarctic and Southern Ocean Coalition (ASOC), congratulated the French government and its licensed fishers for continuing to enforce and monitor fishing procedures that lowered bycatch in the Indian Ocean. By then, seabird mortality in French waters was no higher than in other CCAMLR areas.[30] Part of the improvement resulted from a 2005 treaty between Australia and France that addressed the need to collaborate to ensure responsible fishing around their southern possessions. The joint surveillance aimed at stamping out illegal fishing, and collaboration on marine research and conservation outlined in this bilateral treaty added a new layer of official cooperation. It contained provisions for checking for unlicensed toothfish longliners active in island waters and consultation on a biennial basis to update and analyze data on illegal fishing. The treaty protects the sustainability of both toothfish stocks and seabird populations.[31]

In South Africa, scientists have headed efforts to solve fishery and seabird losses in the exclusive economic zone off the mainland and around islands where five albatross species nest. The BirdLife Marine Programme, operated by BirdLife South Africa, is working to expand fishers' awareness as well as establish effective mitigation regulations. Thirteen species of seabirds currently listed as Endangered, including the Black-browed, Atlantic Yellow-nosed, Indian Yellow-nosed, and Tristan Albatrosses, feed in the fish-rich Benguela Current, and many fall victim to longlines. In addition, South Africa is a signatory range state for fifteen of the twenty-eight species of albatrosses and petrels listed under the 2001 Agreement for the Conservation of Albatrosses and Petrels (ACAP). Claimed by South Africa in late 1947, the Prince Edward Islands are important breeding sites for nine of these fifteen species, most of which are threatened with extinction.

South Africa completed its National Plan of Action to reduce seabird deaths in 2008; it listed ten albatrosses among twenty-six seabird species that were taken in longline fisheries. The mitigation measures for albatrosses and other seabirds follow the accepted FAO combination of fishing devices and procedures, including the use of streamers, weighted lines, and night setting. However, observers have discovered that crews in the deep-sea fleet neither comply with these regulations nor expect them to be enforced. Fishers argue that they were not included in developing the regulations and that instructions are vague and confusing. BirdLife South Africa has suggested that open forums could more effectively solicit feedback as well as provide a neutral space for mutual exchange of observations and information among conservationists and crewmembers. Respect for fishers and the fishing industry and their expertise is essential to obtaining their cooperation in the use of effective bycatch mitigation procedures.

Onboard observations by BirdLife/RSPB Albatross Task Force team members enhance collaboration with fishers by allowing the former to understand how crews make their living in this fish- and bird-rich zone. New Zealand fishers working through the nongovernmental conservation organization Southern Seabird Solutions have assisted counterparts in South Africa in monitoring albatross losses in longline fishing while getting to know vessels and crews and the requirements and challenges of working in their fisheries.[32]

In 2001, Japan threw its hat into the ring by addressing seabird bycatch with its own National Plan of Action—Seabirds, revised in 2005. It appears to date that this is an example of a formal agreement that masks an absence of action. Japan claimed that it would bring "full implementation of mitigation measures"[33] to bear in 2015. It acknowledged that nine species of albatrosses and six species of petrels "may have been subjected to incidental catch"[34] in Japanese tuna longlining in the Southern Ocean and that all three Pacific Ocean albatrosses were subject to incidental take, although the nation did not report taking any. The Japanese plan emphasized educational outreach to fishers and their communities and discussed improvements in *tori* for smaller vessels in coastal waters. The plan reports that vessels fishing for bluefin tuna south of 20°N in the Pacific and any foreign fishers who ply Japanese waters are now required to fly *tori* and adds that additional practices, such as night setting, use of thawed baits, release of live birds, and retention of offal onboard "will be taken."[35] Around the island of Torishima, where Japan's endemic Short-tailed Albatross nests, the plan calls for fishers to "take extra care during breeding season"[36] and for the improvement of erosion controls to protect nesting habitat. Japan has expressed a desire to communicate and cooperate with conservation bodies on both bilateral and regional ways of reducing bird bycatch, notably with CCAMLR and the tuna commissions to which Japan is a signatory. How Japan follows through on these intentions remains to be seen. In sum, Japan's action plan is a step forward for a stakeholder who has a long history of being single-minded in maximizing fishing yield, particularly of tuna species. This official expression of concern for other animals in the ecosystem is a positive development.

Taiwan adopted a National Plan of Action in 2006 that required streamers to be flown by longline vessels, mostly going after tuna, south of 30°S latitude, and reported at the 2007 CCSBT meeting in Tokyo that it would implement the requirement that year along CCSBT guidelines. At the same meeting, Korea reported that it, too, was drawing up a national plan of action for seabirds. How realistic such declarations are on the part of these major fisher nations remains to be seen.

At FAO's Committee on Fisheries meeting in Rome in 2007, BirdLife International suggested an update of best-practice fishing and bycatch mitigation methods for the different types of fisheries to support nations in their efforts to respond to the FAO's international seabird protection blueprint. Two years later in Rome, the same committee acknowledged that two-thirds of members recognized a need for plans of action regarding seabirds and also IUU-related issues; however, money was a constraint as was implementation. Merely 38 percent of respondents had examined the problem of incidental seabird take. Some overall progress could be determined, however, in that the number of national action plans for seabirds had more than doubled in the five years since 2005.[37]

To date, Australia, Brazil, Canada, Ecuador, Peru, the Falkland Islands, Japan, New Zealand, Taiwan, Uruguay, South Africa, and the United States have completed a National Plan of Action—Seabirds. Other nations claim to have a draft or to be drafting a plan. Argentina adopted its national plan in 2010 after nine years of technical discussions and elaborations. Other stakeholders include the European Community, Angola, Norway, Brazil, Chile, the United Kingdom, France, Spain, Namibia, and Vietnam. Experts are pressuring additional nations with longline fleets reported to have caused seabird mortality to draw up plans, including Colombia, Madagascar, Mozambique, Mexico, Panama, Peru, Russia, South Korea, Venezuela, and Ukraine.

CHAPTER 19

Nongovernmentals

GLOBAL NONPROFIT ORGANIZATIONS play a counterbalancing role in addressing some of the negatives of worldwide global markets, including environmental damage and destruction. The role many play derives from a constituency that transcends states' boundaries and political sovereignties. NGOs support research; provide data to national governments and agencies as well as to likeminded international groups; sponsor and cosponsor symposia and conservation projects; publicize the status of animal and plant species; and lobby national and international policy and law makers to enact, implement, and enforce environmentally sound legislation. Under the umbrella of saving threatened and endangered species, habitats, and ecosystems, many NGOs today include threatened seabirds among their concerns. They often provide the most independent and disinterested data available, data on which national governments and international agencies such as fishery and seabird conservation bodies can rely. As fishery bodies have expanded their purview to consider sustainability instead of maximum yield and ecosystems instead of single species, they have increasingly turned to nongovernmental data as a basis for planning and decision making.

More than a dozen NGOs are specifically or primarily involved with albatrosses. The most active of them are based or have offices in the twenty-five albatross range states listed in the 2001 Agreement for the Conservation of Albatrosses and Petrels, but some are not. For example, German bird groups and organizations in the Netherlands, which has a strong tradition of ornithological research in Europe and Africa, have made themselves heard about albatross deaths. Nonbird groups, too, have taken up the albatross cause. In 2003, the US Humane Society alerted members to the environmental costs of bycatch and promoted, so far unsuccessfully, inclusion of the Patagonian Toothfish in the 1975 Convention on International Trade in Endangered Species of Wild Fauna and Flora (CITES).

Public and private universities and museums form a separate category of nongovernmental organizations. Their research faculties provide a body of theoretically disinterested and sometimes conflicting marine science that is an essential component of our knowledge and an invaluable base for decision making in regard to the environment. It was the American Museum of Natural History and the Museum of the Brooklyn Institute of Arts and Sciences that sponsored Murphy's and Alexander's research trips to South Georgia in the early 1900s. Lance Tickell conducted his baseline research on South Georgia under the auspices of Johns Hopkins University. The way nongovernmental advocacy groups interact with and utilize—or fail to utilize—academic research-based knowledge is an interesting story in itself.

The first nongovernmental advocacy organization that took a local, bottom-up stance for protection of the albatross was the New South Wales Albatross Study Group that Doug Gibson and his friends formed in 1951 when they began their albatross bird-banding program. That organization evolved into the current Southern Oceans Seabird Study Association (SOSSA). The Otago Peninsula Trust on New Zealand's South Island formed in 1967 to support Lance Richdale's work to protect the Southern Royal Albatross colony on Taiaroa Head. Additional local and national organizations, many of which also partner with BirdLife International, include the Australian Marine Conservation Society; Birds Australia; the Humane Society International; New Zealand's TerraNature Trust and its Royal Forest and Bird Protection Society; the UK's Royal Society for the Protection of Birds; the US National Audubon Society and Pew Charitable Trusts, which funded Graham Robertson's research on line-weighting for bycatch mitigation; and the American Bird Conservancy (ABC), which has campaigned vigorously for an end to killing seabirds on longlines. Joined by forty other groups, including National Audubon, National Wildlife Federation, and Defenders of Wildlife, ABC lobbies the US National Marine Fisheries Service to mandate practical measures for lowering bird deaths on all its longline vessels and to work with the additional thirty or so nations and territories with longline fleets to draw up and improve national plans of action to minimize seabird drowning. It is a comprehensive and assertive program.[1]

International NGOs that have focused on albatross protection efforts include the World Fund for Nature, Greenpeace International, and the World Conservation Union (IUCN). The most influential and effective international organization in regard to albatross conservation has been BirdLife International and its global partners. Southern Seabird Solutions, formed originally under the umbrella of the New Zealand Department of Conservation and Ministry of Fisheries, has also figured in efforts to save the albatross.

After BirdLife introduced its Global Seabirds Program in 1997, it launched a Save the Albatross campaign at the twelfth British Birdwatching Fair, called Bird-Fair, in 2000. Two bird enthusiasts devised the three-day annual BirdFair in the late 1980s, and now two hundred exhibitors, including bird guides, tour operators, suppliers of birding equipment and clothing, wildlife artists, and bird food producers, gather to represent the entire bird-watching industry. In recent years, the Royal Society for the Protection of Birds and the Leicestershire & Rutland Wildlife Trust have organized BirdFair at Anglian Water's Rutland Water Nature Reserve in Egleton, luring thousands of people into its bird-watching center.

The year before, BirdFair had raised funds for Brazil's Atlantic forest birds, and when BirdLife proposed its Save the Albatross campaign in 2000, BirdFair agreed to back it. A first step was to let birders, the interested public, and the media know about perils from longlining. Consumers in particular, organizers believed, would be able to bring political pressure on fisheries and the industry once they knew about the kinds of fishing that killed birds. The alliance between BirdLife and BirdFair has been successful, bringing to public awareness the hardships faced by albatrosses and stimulating global citizen outreach. BirdFair also raised almost US$250,000 for albatross conservation.[2]

Under its Important Bird Areas (IBA) Program, begun in the 1980s, Bird-Life in 2005 began using its considerable scientific expertise to begin identifying Marine IBAs. There are 9,000 mostly terrestrial and freshwater sites listed worldwide as critical spaces for bird species, and today there are just over 2,100 Marine IBAs. Several have been designated for pelagic species, and more will follow. With the leadership of Deon Nel and John Croxall, and subsequently under the direction of Cleo Small, BirdLife produced the 2004 *Tracking Ocean Wanderers*, which graphically portrays Marine Important Bird Areas for albatrosses. Using banding data and telemetry, this pathfinding book maps, through stunning visual displays and explanatory text, the at-sea distributions of albatrosses and petrels and their overlap with major fisheries. Marine Important Bird Areas for albatrosses include the so-called Barrow to Baja Project, which identifies bird-rich areas along the northwest Pacific coast, the Humboldt Current of western South

America, the Patagonian Shelf off Argentina, the Antarctic Polar Frontal Zone north of the Antarctic continent, and the Benguela Current off southern Africa. Newly developed criteria worked out and standardized by Spanish experts also identify coastal and inshore locations where seabirds congregate to rest, molt, or forage from nearby nest colonies.[3]

Ecuador became the first Southern Hemisphere nation to officially designate Marine Important Bird Areas. Ecuador has more than sixteen hundred bird species in an area the size of the United Kingdom, and its bird experts have identified and worked to protect critical habitat in more than one hundred Important Bird Areas. The list includes ten sites on the Galápagos Islands, where there is concern about the effects of longlining on the islands' ten thousand pairs of breeding Waved Albatrosses. Ecuador has designated an additional forty-seven Marine Important Bird Areas along its mainland coast. Aves & Conservación (BirdLife Partner for Ecuador), formerly CECIA, works with government officials on developing and supporting the IBA program.[4]

BirdLife partners Spain and Portugal have also listed IBAs and Marine IBAs in and around the European Union. BirdLife's New Zealand partner Royal Forest and Bird has designated similar conservation areas in the South Pacific region. Currently, sea areas adjacent to World Heritage Sites are off-limits to fishing and thus protected areas for birds. In addition, the 2003 Auckland Islands Marine Park, totaling over 1.2 million acres, is a precursor for additional sea reserves off that nation's islands. All impose limits for entry and prohibit most fishing. Efforts are also under way to declare as much as 20 percent of New Zealand's marine exclusive economic zone a Marine IBA and reserve. This move would benefit the nearly one-fourth of the world's seabird species that breed in New Zealand. Other island reserves claimed by the United Kingdom and South Africa exclude fishing inshore and are de facto marine sanctuaries. It is not just New Zealand that has little of its sea area under protection. Less than 1 percent of the entire planet's marine mantle is protected, a fact not lost on the Convention on Biological Diversity. This body is moving to amend that situation, and BirdLife and RSPB's transponder tracking data is essential information for identifying hot spots for seabird convergence, which happens in areas far removed from coastlines and human presence.

In October 2005, BirdLife began Operation Ocean Task Force, now the Albatross Task Force, an ambitious collaboration with the Royal Society for the Protection of Birds to develop an international team of mitigation instructors to work with crews on longline vessels on techniques that decrease seabird bycatch in their fisheries. BirdLife representatives speculated at the time that an albatross was dying every five minutes from a longline hook. By the close of 2007, the Albatross Task Force had eight instructors working aboard vessels off South Africa, Brazil, and Chile. In Chile, the instructors had the advantage of working under Carlos Moreno, who had convened the first meeting of CCAMLR's bycatch working group. Along with Rodrigo Hucke-Gaete, he founded a Chilean nongovernmental organization called Centro Ballena Azul to collaborate closely with both fishers and the central government to improve fishing practices and seabird conservation. Two years later, there were fifteen Albatross Task Force instructors working in seven southern African and South American nations. Their aim is to connect fishers' experience at sea with the devices and techniques that improve awareness about conserving and saving seabirds. It is important to fishers to be able to reduce seabird deaths without being blamed for reduced catch or for the imposition of new gear and ways that challenge habitual ways of fishing.[5]

This effort to reduce seabird deaths by training crews worldwide is one component of BirdLife's Global Seabirds Program. Another provides leadership and support to local conservationists and stakeholders to increase their effectiveness in working with their own fishers and government agencies to reduce albatross losses. In South America in 2001, with backing from Global Seabirds and other nongovernmental organizations, Aves Uruguay with SEO/BirdLife Spain's Carlos Carboneras organized a workshop for representatives from nations with fisheries in the southwest Atlantic Ocean—Brazil, Uruguay, and Argentina—and in the southeast Pacific Ocean—Chile, Peru, and Ecuador. At this first workshop in Punta del Este, Uruguay, Ben Sullivan, formerly of Falklands Conservation but by then director of Global Seabirds, and Deon Nel, then with BirdLife South Africa, described the problem and suggested conservation priorities. They discussed the newly crafted Agreement for the Conservation of Albatrosses and Petrels and the need for South American countries to formulate national plans of action for seabirds.

Participants at the Punta del Este meeting agreed that albatrosses and other seabirds were dying off South America. Reportedly twelve thousand birds were lost annually in pelagic longlining for swordfish out of Brazil's port of Recife. Under Tatiana Neves's Projeto Albatroz, a number of fishers had agreed to take measures to lower that take of seabirds while the Brazilian government was formulating its national plan of action. This included deploying locally made streamers and dyeing baits blue. From Argentina, fishers reported snagging only one thousand seabirds from the almost 22 million hooks they set per season off Patagonia, mostly Black-browed Albatrosses and White-chinned Petrels. If these numbers were correct, the low bird-hook ratio of .001 birds per 1,000 baited hooks is unequalled in longline fishing. However, a fledgling Argentinian on-board observer program reported losses five or six times greater in that area. Workshop participants called for increaing the number of observers in order to generate better data.

The workshop concluded with a fourteen-point agreement, the South American Strategy for the Conservation of Albatrosses and Petrels (ESCAPE), under which each national group would develop effective mitigation actions while continuing to fish. It also included improving devices and procedures, training and educating skippers and their crews and the general public about longlining, and lobbying with agencies and politicians for governmental actions to protect seabirds.[6]

A second regional workshop, in Valdivia, Chile, in December 2003, revisited earlier issues. It was only several months before the second workshop that several South American delegations had complained to the FAO Committee on Fisheries that FAO's International Plan of Action—Seabirds had taken little or no account of bird losses in South American waters. Importantly, the plan also failed to provide South American countries the technical expertise and assistance needed to implement effective bycatch mitigation. In response, the FAO initiated the Valdivia workshop in collaboration with BirdLife to address South American concerns. The twenty-six participants, who included representatives from five South American nations, Kim Rivera from the fisheries division of the US National Oceanic and Atmospheric Administration, Carlos Carboneras, and New Zealander Janice Molloy representing Southern Seabird Solutions, revisited the topic of seabirds and longlining. New or preliminary action plan reports from Argentina, Chile, and Brazil; a round table discussion of mitigation techniques; and agreement on priorities and programs moved this workshop from the primary emphasis at the Uruguay meeting on the need for scientific data into a potentially valuable collaboration among South American stakeholders in fish and albatross survival.

Carlos Carboneras updated participants on albatross conservation. Since the Uruguay meeting, Ecuador had signed and ratified the Agreement for the Conservation of Albatrosses and Petrels. The Falklands/Malvinas action plan was in draft stage, and Brazil and Argentina were about to put pen to paper. Carboneras also reported that BirdLife International had helped the 1969 International Commission for the Conservation of Atlantic Tuna, the earliest of the fishery management bodies with significance for albatrosses, to formulate and adopt its first resolution calling for reduction of seabird bycatch. Additionally, there had been a conference on illegal fishing in Spain, and the Coalition of Legal

Toothfish Operators was actively exposing illegal toothfish vessels in the Southern Hemisphere. Carboneras applauded the momentum gained at the Uruguay meeting, and commended ESCAPE for pressing for an ecosystem approach in South American fishery management. He emphasized that the purpose of the workshop was to assist each participating nation in drawing up and implementing an action plan for seabirds.

The Valdivia workshop recommended the standardization of bycatch data and announced the goal of making seabird populations and fish stocks sustainable through sound management, but the conferees insisted that a balance had to be struck between saving the livelihoods of fishers and the lives of wild fish and albatrosses. Fishers and environmentalists needed to negotiate a trade-off, with special attention to artisan stakeholders who had little experience in protecting seabirds. Peru's recently expanded inshore artisan fleet was busily snagging albatrosses, for example, and needed to be brought onboard, so to speak, in all discussions about reducing losses.[7]

In September 2005, BirdLife International's research revealed still alarming decreases in the populations of six of the twenty-two albatross species, including one previously considered safe, despite the multilevel and multifocus global work BirdLife was doing to save seabirds. Five years later, BirdLife International significantly increased the threat status of albatrosses (see Table 5), and longline fishers also reduced the Spectacled Petrel to a vulnerable state, with other species of its kind heading in that direction.

In July 2002, Janice Molloy, at that time a biologist in New Zealand's Department of Conservation, organized and convened a workshop in Nelson, the result of collaboration between her department and the Ministry of Fisheries. The outcome, Southern Seabird Solutions, was an alliance of government, nongovernmental, Maori, and fishing industry groups to promote mitigation of seabird bycatch. Based in Wellington, Southern Seabird Solutions' work encompasses the entire Southern Ocean where seabirds roam. Incorporated as a trust in November 2003, it imposes a levy on the fishing industry to support education and research. Its objectives are closely aligned with those outlined in the 2001 International Agreement on Albatrosses and Petrels, and its members include local, indigenous, and commercial fishers and fishery managers; government agencies; nongovernmental organizations; Antarctic tour operators; and seabird ecologists. Because it is fishers themselves who hold the key to solving the problem, Southern Seabird Solutions, like BirdLife's Albatross Task Force, has worked to support New Zealand fishers in becoming role models and leaders worldwide in regard to best fishing practices and mitigation techniques. It has done a good job. Southern Seabird Solutions convenes meetings and conferences to exchange information with fishers and environmentalists, like the South American Fishers Forum it sponsored in Brazil in December 2006. Sixty-five participants from seven nations gathered to discuss bycatch mitigation. A Southern Seabird Solutions video, *Fishing the Seabird Smart Way*, demonstrated the newest technologies and procedures. It appeared that many of the commercial operators at the meeting already had solid knowledge about the problem and its mitigation, but artisan fishers welcomed assistance in developing types of mitigation suited to their fishing.[8]

Recent projects include the annual Seabird Bycatch Avoidance Award to a New Zealand fishing owner or crew that has taken effective steps to reduce bird losses. An exchange program is in place whereby Southern Seabird personnel travel to swap information about mitigation methods with fishers throughout the Southern Ocean. In 2003, the first exchange placed a commercial swordfish skipper from Chile on a New Zealand longliner vessel. Management committee member Malcolm McNeill made a five-day visit in August 2004 to France's Reunion Island to demonstrate Kiwi-built bird-scaring lines to skippers from France's seven-vessel Patagonian Toothfish fleet, and a French skipper followed up with a visit to New Zealand fisheries. Southern Seabird Solutions has also helped a South African alliance organize its own seabird solutions group.[9]

A 2005 cooperation agreement between Southern Seabird Solutions and the Instituto del Mar del Perú (IMARPE), Peru's Institute of Fisheries Research, to promote fishing practices that reduced seabird bycatch in that nation's waters was a major achievement. New Zealand's Critically Endangered Chatham Albatross forages in convergence waters off Peru, so procedures to lower catches of birds benefit both nations. A few months later, with funding from the New Zealand Ministry of Foreign Affairs and Trade and the Department of Conservation, the International Association of Antarctic Tour Operators, and IMARPE, Southern Seabird Solutions sponsored a trip to Peru for fisher Dave Kellian, inventor and advocate for sustainable fishing. Kellian has committed his own time and money to developing new solutions like the underwater bait-setting capsule he designed and bird-scaring lines for the smaller vessels, which use lower line-setting speeds. Because of his experience in designing and testing the capsule, Kellian was one of thirty-eight people nominated in the sustainable business category of the New Zealand Ministry for the Environment's annual Green Ribbon Award in 2004. In Peru, Kellian worked for three weeks alongside government officials and IMARPE and nongovernmental Pro Dephinos staff to identify the kinds of support needed to address seabird bycatch in Peru.[10]

Southern Seabird Solutions is a unique, hands-on collaborative. It provides space for open dialogue between the key stakeholders, those most able to define and protect their own interests. Participants have been able to develop and leverage linked issues, conduct negotiations, and make trade-offs that accommodate the interests of all. Obtaining the participation of those most knowledgeable about what actually happens onboard and about the actual, quantified risk to seabird species from commercial fishing creates the optimum conditions for stopping what is both economically and biologically wasteful. Southern Seabird Solutions makes available information about seabird losses on both a species and a regional basis. It keeps abreast of best fishing practices and presents intelligent choices to fishers, wholesalers, and consumers. Southern Seabird Solutions recognizes that "New Zealand is the 'seabird capital' of the world. Our waters and offshore islands are the breeding grounds for more albatross and petrel species than any other country. But, while they breed here, many of these birds live much of their lives elsewhere . . . If we don't change the way we fish, many of these species will disappear."[11]

CHAPTER 20

Trade

COMMERCIAL FISHERS FROM nine nations make up the Coalition of Legal Toothfish Operators (COLTO), formed in 2000 to fight illegal fishing. Twenty-three or so COLTO members operate on the basis of sustainability and support catch quotas and seasonal and area closures. The group does not address bycatch, but its members operating in the Southern Ocean cooperate with CCAMLR in regard to best fishing practices and adopt seabird avoidance measures. Some members are also working with the Marine Stewardship Council to meet the criteria for ecolabeling their catches.

Legal toothfish fishing takes place in the high seas of the Antarctic and sub-Antarctic and in several Southern Ocean exclusive economic zones. COLTO fishers harvest about fifteen thousand tons of toothfish annually, close to half of the world's annual legal catch. Illegal fishers, however, also operate in the same waters, so COLTO members work together to provide surveillance and other valuable information to governments and the public at large to help stop the toothfish poachers. COLTO publishes information about illegal operations by owners, flag states, vessels, and crews on its website and in correspondence and publications. It also reports their fates.[1] In 2007, COLTO named thirty-five vessels known to be hunting Patagonian Toothfish illegally, including the infamous *Viarsa 1*. Two had been forfeited, one was convicted of illegal fishing, and one was scuttled to make an artificial reef. Recent worldwide concerns about illegal fishing in fisheries other than toothfish fisheries have also engaged COLTO efforts. As a result, COLTO collaborates with CCAMLR to report miscreants who operate out of ports in Namibia, Indonesia, Singapore, Hong Kong, and until re- cently in Mauritius. COLTO has also publicized the consortiums to which some illegal operators belong.[2]

In 2000, the Organization for the Promotion of Responsible Tuna Fisher-ies (OPRT) was organized with members from Japan, six other Asian nations, and Ecuador. Based in Tokyo, its membership welcomes all stakeholders in tuna fishing and fisheries who agree to take responsibility for sustainable ecosystems. Illegal fishing is one of its focuses. It monitors trade information from the Japa-nese market and buys back vessels caught fishing illegally to render them for scrap. In 2005, OPRT hosted the third of four International Fishers Forums to date to exchange ideas and information on measures being used or developed to reduce capture of seabirds on longlines.[3]

In 2003, the Organization for Economic Cooperation and Development's Round Table on Sustainable Development (RTSD) focused on how to minimize illegal fishing on the high seas and provide for orderly management of high-seas fisheries. One of the outcomes of the round-table discussions was the formation of a High Seas Task Force to study illegal fishing and develop a plan for eliminat-ing it. As the world's international development agency for private enterprise, such a study is well within OECD's purview. OECD took over from the 1961 Or-ganization for European Economic Cooperation (OEEC), formed to administer American and Canadian aid under the Marshall Plan for reconstruction of Eu-rope after World War II. It works to strengthen the economies of its member na-tions, both industrialized and developing, in a global market. The Round Table on Sustainable Development was established in 1998 to provide a setting for an ongoing discussion between a highly diverse international group of ministers and business and civil society leaders of specific topics across the spectrum of economic and environmental sustainability.[4]

The High Seas Task Force includes fisheries ministers from Australia, Cana-da, Chile, Namibia, New Zealand, and the United Kingdom, and representatives of three international nongovernmental organizations: the World Conservation Union (IUCN), the World Wildlife Fund, and the Earth Institute at Columbia University in New York. In March 2006, the task force released its first report on illegal fishing, titled *Closing the Net: Stopping Illegal Fishing on the High Seas*, with recommendations that closely align with and reinforce the principles of the Fish Stocks Agreement and the Code of Conduct for Responsible Fisheries. It rec-ommended, first, a two-year survey of illegal fishing and its catch and bycatch. Bolstering previous issues taken up by CCAMLR and the FAO's Committee of Fisheries, the officials pointed to six action areas. It argued that international multilateral accords about fisheries surveillance, flag-state performance, and port management were essential to any effort to control illegal fishing. It also recommended requiring global positioning systems on all high-seas fishing vessels, especially necessary to stop toothfish piracy, and additional port and trade measures to help close markets to illegally taken fish.

The report made two assessments with regard to seabird mortality. First, it estimated that illegal fishing may be increasing losses of albatrosses and petrels by as much as 36 percent over the number of bird deaths from natural conditions and legal fishing. Second, while the report acknowledged that illegal fishers likely do not deploy bird-scaring lines or other bycatch mitigation devices, it called forcefully for the use of these and other mitigation methods by all legitimate longline fishers. Given the limited number of onboard observers to document the employment of mitigation devices and monitor bycatch, and "the lack of reporting of catch of target and bycatch species, their size, composition, and where and when such catches are made,"[5] the ongoing situation in regard to bycatch reduction would be hard to determine. High Seas Task Force officials planned to establish local and area networks among the OECD membership that would continue to monitor and collect comprehensive data about catch and bycatch of legal fishers and extrapolate the results to calculate a best estimate of catch and bycatch of illegal operators. This would provide a baseline for measuring progress in reducing illegal fishing globally. The United Kingdom directly supports the work of the High Seas Task Force. It has established a commission in the Department for International Development to conduct an analysis of illegal fishing that includes economic, social, environmental, ecological, health, and nutritional costs.

In 2004, five years after the International Plan of Action—Seabirds was published and while the High Seas Task Force report was in final review, marine expert Helen Bours of the Environmental Justice Foundation asked, "What's left to discuss?"[6] The Environmental Justice Foundation has collaborated with Greenpeace in actively thwarting illegal fishing by sailing vessels into areas where illegal fishers are operating at a cost of up to US$9 billion per year. Three years later, Bours still wanted to know when governments would start acting. "These vessels are at sea for years," she says. "They transfer their fish onto other vessels, they get refueled at sea, even the crews are changed at sea. . . . It's another world."[7] That is the challenge for all international environmental regimes, to mobilize the political will to do what has to be done in organizations that often work on a consensus basis and frame issues in ways that don't offend the sensibilities of nations, the regional fisheries bodies that manage the resources, or the commercial interests that harvest them.

The US Western Pacific Regional Fishery Management Council convened the Second International Fishers Forum in Honolulu in 2002. That second meeting was attended by a greater diversity of fishery participants from around the world than the first Forum, with 236 participants from twenty-eight different countries. It focused not only on seabird longline bycatch but also on incidental catches of sea turtles in pelagic longline fisheries, mitigation of which it added to its mission statement. In July 2005, tuna fishers gathered in Yokohama, Japan, from all over the world for an International Tuna Fishers Conference to meet jointly with the Third International Fishers Forum, which was sponsored by the Organization for the Promotion of Responsible Tuna Fisheries.

Two of the thirteen points in the Yokohama Declaration the participants issued after the meeting represented advances for seabirds. Tuna operators from all oceans agreed to "refrain from circumventing conservation and management measures by re-flagging our fishing vessels to nonparties of relevant [fishery management bodies] with little or no ability to control their fishing vessels."[8] Second, participants agreed that approaching fisheries as parts of interrelated ecosystems was necessary to sustain a healthy ocean and promised to vigorously promote the use of fishing techniques that reduce seabird bycatch. Participants, agreeing with the approach to mitigation taken by the Western and Central Pacific Tuna Fishing Commission, called for tailoring mitigation techniques to specific tuna fisheries and ensuring crew safety.

The Fourth International Fishers Forum took place in Costa Rica in November 2007, backed by Japan, the US Western Pacific Regional Fishery Management Council, and the Costa Rica Fisheries and Aquaculture Institute (INCOPESCA) with assistance from the local World Wildlife Fund office. At this meeting, fishers added bycatch management for two additional species, sharks and cetaceans, and the goal of developing bycatch avoidance strategies through viable and equitable regulatory methods. The overall aim of these International Fishers Forums remains that of promoting responsible longline fishing.[9]

The International Association of Antarctic Tour Operators, begun with the assistance of Graham Robertson in 1999, continues to generate interest in the Southern Ocean environment and ecology among its guests and contributes a percentage of its profits to albatross conservation. Tourist numbers in the Antarctic have increased eighteenfold since the 1950s. From five thousand per year in the 1990s, it climbed to thirty thousand per year in the first decade of the twenty-first century, mainly on the Antarctic Peninsula. In 2001, IAATO began providing lectures and films to its guests about fisheries and seabird bycatch and soliciting contributions for albatross conservation. Birds Australia directs IAATO contributions toward needed research. In IAATO's first Seabird

Newsletter, Graham Robertson listed sixteen research and training projects benefiting from tour contributions. The first project, awarded a grant of $5,000, was New Zealand fisher Malcolm McNeill's experiments with fast-sinking longlines. IAATO contributions grew from $5,000 to $75,000 between 2002 and 2005, and in 2006, funds from eighty outfitters supported bycatch mitigation in South American waters, specifically off Uruguay, Peru, and Chile. The International Polar Year from March 2007 through March 2009, the fourth since the first one in 1882–1883, adds the possibility for a new tier of studies. It also increases the number of tourists to the Southern Ocean, who are increasingly knowledgeable about issues of climate, environment, and ecology. After experiencing that region of the world for themselves, they are likely to realize they are also stakeholders in the health of the Southern Ocean and Antarctica.[10]

CHAPTER 21

Celebrities

AT THE TURN OF THE CENTURY, the albatross began to acquire a certain amount of fame. The 2002–2003 America's Cup yacht race, held in New Zealand, threw its support behind the albatross, and Forest & Bird, which had launched its Save the Albatross campaign in partnership with BirdLife International, declared "Yachties Support Albatrosses."[1] At a press conference in Auckland, representatives of America's Cup syndicates Team New Zealand, Alinghi from Italy, OneWorld from the United States, and British Challenge from the United Kingdom signed and mailed postcards to New Zealand officials and to diplomatic missions in seven other range states. In 2006 in Wellington, six skippers from the Volvo Ocean Race teams pledged their support to albatrosses. "It can get pretty lonely when you're at sea for weeks on end, so seeing these awesome birds is a great sight for us,"[2] said Mike Sanderson, skipper of *ABN Amro One*. And he noted that he was seeing fewer and fewer albatrosses at sea.

Early in 2004, British TV personality and nature specialist David Bellamy announced a Big Bird Race, promoting it as a great long-distance wager. Ladbrokes, the British bookmaking conglomerate, underwrote it. Punters, the English betting public who were used to betting on horses and greyhounds, were invited to wager on which of eighteen electronically tagged young Tasmanian Shy Albatrosses would be first to race over six thousand miles of open ocean. The plan was for them to leave their nest sites on one of three small rocky islands off the coast of Tasmania and fly to the coastal waters off South Africa. Wagerers, it was announced, could expect their selections to glide and soar on a migration lasting four to six months across the Indian Ocean. The winner would be the first to pass the finish line at 31°E longitude.

The Big Bird Race aimed to raise money for albatrosses and bring their plight to the attention of national and international audiences. The London-based Conservation Foundation thought up the idea, and albatross biologists like Rosemary Gales in Tasmania selected and tagged the young entrants. Famous people bought or sponsored individual birds. Queen Noor of Jordan, President Emeritus of BirdLife International, sponsored Ancient Mariner. Brian May, the former guitarist for Queen, the English rock band formed in the 1970s, sponsored Rocky, and model Jerry Hall, former wife of Mick Jagger, named her selection Aphrodite. Other racers were Ocean Spirit, Ecologist, Styx, and Paragon.

Unfortunately, some youngsters died in the nest, such as Lucky Seven, sponsored by champion flat-race jockey Frankie Dettori. Others may have decided not to migrate after becoming confused by fierce gales during fledging. By July 1, two months into their flight westward, fifteen of the eighteen birds were missing. Some commentators presumed them dead, and others speculated that the electronic tags were faulty. Aphrodite went silent for two weeks. One albatross, Harriet, flew east before turning north away from South Africa. Queen Noor's Ancient Mariner last signaled on May 29, 2004, apparently having reversed its direction. Hall's Aphrodite then beeped into life and took the lead. Although hopes about bird survival and completion of the race were dimming, Ladbrokes reported 2,500 bets on the Big Bird Race, about US$20,000 for albatross protection and conservation. Hall's Aphrodite won. Listeners picked up her signal in South African waters on July 12 and declared her first past the post, or "official winner." Two more birds crossed the finish line a little later. Prince Philip presented Hall with the trophy he had donated for the Big Bird Race, the Duke of Edinburgh's Challenge Cup.[3]

In May 2005, a second Big Bird Race began. Backers included the previous winner, Jerry Hall, joined by Sir David Attenborough, Olivia Newton-John, and fourteen other celebrities and representatives of animal groups. Ladbrokes funded satellite transmitters for twenty birds, and Gales and others helped tag eighteen young birds. The race began on May 2, 2005, but within a month, disaster struck. "There has been unprecedented mortality amongst juvenile Tasmanian Shy Albatrosses,"[4] declared Tim Nevard, Conservation Foundation organizer and commentator. All transmitters had fallen silent. The radioed birds had apparently died in Tasmanian or Commonwealth shelf waters. Compared with the relative success of the first race, when at least three birds completed the journey, the 2005 race was a sad debacle. However, it could not more effectively

234

have made the point. Albatrosses lead risky lives with high rates of attrition. It was easy to see that the entire family might be endangered. The races and their celebrity punters raised interest in the albatross and dramatized to people in all walks of life the downward spiral for these seabird populations.

Prince Charles became familiar with seabirds and their grace and mystery during his service in the Southern Ocean as a Royal Naval officer. He recalls the unforgettable sight of an albatross gliding for hour after hour alongside his ship, maintaining perfect position all that time. He began publicizing his personal commitment to the conservation of the albatross and his concern over its plight resulting from longline fishing in an article he wrote for *The Field* in February 2001. He noted that barely eight years earlier three of the twenty-one species were officially classed as threatened, but that by 2000, as he sat down to write the article, that number had risen to sixteen. By 2004 it would rise to nineteen. Prince Charles has been a consistent advocate for their protection, expressing personal disquiet to staff of BirdLife and its UK partner, the Royal Society for the Protection of Birds. In 2002, he introduced BirdLife's Save the Albatross campaign to an audience of invited guests at St. James's Palace, characterizing the situation for the seabirds as "dreadful . . . and an entirely uneccessary one."[5]

In September of that year, when Prince Charles welcomed delegates to the Convention for Migratory Species of Wild Animals at the Seventh Conference of the Parties in Bonn, he referred to the Agreement on the Conservation of Albatrosses and Petrels, open for signatures, which had developed under the leadership of Australia and South Africa. He called on the world community, including range-state governments and those with fishing fleets in albatross waters, to ratify and get the agreement working to stop the albatross's decline toward extinction.[6]

At that time, eight nations had signed and two had ratified the agreement. In 2004 at the first meeting of ACAP parties in Hobart, the chair read a letter from Prince Charles expressing his strong support and best wishes for the work ahead. His Royal Highness correctly places albatross declines in the broader context of fisheries, the fishing industry, and the marine world, characterizing as shallow or "empty" the rhetoric about sustainable development when it applies to fisheries. He argues that the albatross's problem painfully exposes the superficiality of claims of sustainability in the high seas, where an unprecedented technological capacity is reorganizing ecosystems in indelible ways. "To me, the albatross may be the ultimate test of whether or not, as a species ourselves, we are serious about conservation: capable of co-existing on this planet with other species,"[7] said the Prince, because there are neither shortcuts nor second chances for saving it.

Prince Charles has challenged conventional conservation methods in regard to the albatross. He declares no marine park or preserve is large enough to protect a single species; no captive breeding program can assist recovery; no single corporation or government sponsor can do more than raise awareness; and no single nation can take unilateral action that is effective. Time is running out for some albatrosses unless collaborative mechanisms can come into play. Even if albatross bycatch were to end today, he notes, given the natural history of the seabirds with their one chick every two years, their population recovery will have to be protected over decades. An "appalling tragedy [is] unfolding, out of sight and out of mind, in the vast foam-flecked spaces of the Southern Ocean,"[8] he warns.

Three years after introducing the Save the Albatross campaign, the Prince rallied support for the Albatross Task Force. During a speech at an RSPB albatross gathering at Trinity House in London, he described his visit to Taiaroa Head to observe the Royal Albatrosses:

> During my visit, I sat literally a foot away from one of these magnificent birds, as it gazed trustingly and quizzically with its extraordinarily endearing features—without fear—at this strange intruder surrounded by people with exceptionally large cameras. It made me think that there is no greater symbol of man's lunacy than our excessive industrialized fishing systems, using lines stretching out eighty miles—I thought it was a typing error some years ago—behind huge fishing boats which hook and drown 100,000 of these mysteriously unafraid and serene birds each year.[9]

Since his New Zealand visit, Prince Charles has given further support to the Television Trust for the Environment's Earth Report, broadcast by the BBC in November 2006. He also recognizes the potential of consumer power in regard to both fish and albatross conservation. In the film *Race to Save the Albatross*, he warns about further losses of the remarkable birds and encourages consumers to be vigilant about what wild fish they purchase, taking home supplies only from certified stocks. He goes on to say, "It calls for a major effort of international co-operation and for regional fisheries to demonstrate sea-bird-friendly fishing of all vessels plying their waters. And there is not much time left."[10]

In late June 2004, sailors John and Marie Christine Ridgway traveled to the UN's FAO Headquarters in Rome to present a petition to save the albatross from pirate fishing with over one hundred thousand signatures from residents in 131 countries. They had just completed a sailing trip around the world to raise awareness for the albatross. Ridgway was already a national hero in the United

Kingdom and, to some extent, worldwide. He was well acquainted with the dangers of ocean travel, having made an Atlantic crossing with Chay Bly in 1966, during which two fellow rowers drowned.

In 2003, Ridgway and his wife decided to sail into the albatross latitudes from their home in Scotland in their thirty-five-year-old, fifty-seven-foot ketch, *English Rose VI*. Ridgway said he could no longer stand by while losses occurred in longlining. Under the banner of the United Nations Environmental Program and with a volunteer crew, the sixty-five-year old salt and his wife headed south on July 27, 2003, from Ardmore, Scotland, stopping in Cape Town via the Canaries, then sailing on south to Kerguelen Island, Melbourne in Australia, and Wellington in New Zealand. By the time Ridgway's *English Rose* left New Zealand on January 24, 2004, after a two-week call, his BirdLife sponsor, New Zealand's Forest & Bird, had collected more than ten thousand signatures to save the albatross. Spectators had given Ridgway and his crew a grand welcome and a fanfare departure. Ridgway nosed his ketch out of Wellington harbor and turned east for the long open leg of big seas across the Pacific, through the Drake Passage below South America, and on to Port Stanley, the capital of the Falkland Islands. Through media appearances in each port, and by means of direct links to the Save the Albatross website, the Ridgways and their crew publicized concern over the plight of the albatross and urged people worldwide to pledge themselves to saving these marine birds.

Almost five thousand nautical miles from Wellington on March 3 and more than twenty thousand miles from his starting point in Scotland, he was approaching the Falkland Islands when he saw two Falkland Island Company launches coming out to greet them. Having been battened down in a fifty-seven-foot boat with a six-member crew for six weeks, he said that "this heartwarming welcome was such a shock that half an hour passed before we had the wit to even call them up on the VHF." The two boats "crowded with friendly people"[11] convoyed the *English Rose* the final six miles into Port Stanley, where they were greeted by the governor.

Ten days later, Ridgway pointed his vessel almost due north in the Atlantic Ocean. He put in at the Azores and finally on July 17, 2004, skipped the *English Rose* under the span of London's Tower Bridge, raised in honor of the doughty seafarer and his wife and mates. RSPB officials and the media greeted them, and at the official reception the following day, a letter from Prince Charles praised as unique the gesture that the Ridgways and their colleagues had made for albatross awareness. He noted also that it was during the Ridgways' year at sea that the United Kingdom had ratified the Agreement for the Conservation of Albatrosses and Petrels (ACAP).

Button fathered the five-hundredth chick, named Toroa, recorded on Taiaroa Head, Dunedin, and was the last male born to "elderly" Grandma in 1989. Hand-fed in the nest, he is a regular member of the New Zealand colony. Courtesy Isobel Burns and Lyndon Perriman.

It's not accidental that many of the celebrities who spread the word about albatrosses are sailors. Dame Ellen MacArthur, named a Dame in 2005 for her record-breaking single-handed sail around the world, had several albatrosses follow her at various points along her journey, keeping her company on what was a very solitary experience. She finds the albatross to be "one of the most amazing birds in existence; it is truly breathtaking."[12] As a result, she has added her name and active support to albatross conservation.

In April 2005, she joined forces with Prince Charles to advocate protection for the birds and later traveled with backing from the Royal Society for the Protection of Birds to South Georgia to assist with the periodic Wandering Albatross census on Bird Island. In a similar manner, Sir David Attenborough, British naturalist, broadcaster, and film personality, and vice-president of the Royal Society for the Protection of Birds, has lent his fame in the United Kingdom and elsewhere to saving the grand seabirds. It was Sir David who narrated the film about Grandma and her son Button, who still thrives, in the Royal Albatross colony on Taiaroa Head. That was just one of his films about albatrosses, about whom he notes, "When mankind first entered their domain, as little as 500 years ago, albatrosses had already been masters of the oceans for 50 million years." He concludes, "My personal concern for the future of these majestic birds is echoed by millions of other people across the world—many of whom may never be lucky enough to see an albatross. Like me, they care passionately that these ancient mariners should be given a fighting chance to spread their wings and enjoy another 50 million years."[13]

CHAPTER 22

Capstone

THE 1979 CONVENTION ON MIGRATORY Species began with a terrestrial focus on European fauna, but it but broadened its purview to migratory animals globally, including those in national waters and the high seas. It was Australia's nomination in 1997 of twelve albatross species to the Appendices of the Convention on Migratory Species that put seabird survival on the CMS agenda. To this issue the parties soon added the high numbers of migratory fish being lost as bycatch. South Africa recommended the addition of petrels, bringing the total list at the Convention's Sixth Conference of Parties to two albatrosses and three petrels in Appendix I and twelve albatrosses and six petrels in Appendix II.

Australia offered to start discussions on reaching an agreement among albatross range states about conserving seabirds, an offer that was accepted by resolution. These discussions created a forum for scientists to make the case that commercial fishing, notably longlining for toothfish and tuna, was seriously jeopardizing seabird populations. With support and encouragement from abroad, Australian officials, Antarctic Division researchers, and Tasmanian scientists drew up, under the umbrella of the Convention on Migratory Species, a draft agreement for the protection of albatrosses and petrels. Stephen Hunter, head of the Biodiversity Group of Environment Australia and provisional chairman of the first meeting to negotiate and establish the framework for the new agreement, described existing binding and nonbinding instruments such as CCAMLR and CCSBT, the Fish Stocks Agreement, the Code of Conduct for Responsible Fisheries, and the FAO's International Plan of Action—Seabirds as useful interlocking support structures for the more specialized regime that he anticipated. The Convention on Biological Diversity and the World Heritage Program also provided useful anchor points for weaving together a textured set of codes and regulations for saving the albatross. In arguing for an integrated and coordinated approach, Hunter invoked the model of the Convention on Migratory Species, which with UN backing had already developed an integrated approach to marine, terrestrial, and species-specific conservation actions. Using this model would eliminate the need for more than one agreement, he argued, and participants agreed to include the chair of the Scientific Council of the Convention on Migratory Species in formulating the final version of ACAP.[1]

The United Kingdom urged drawing also on the structures, measures, and legal support of agreements such as the Convention on Biological Diversity, CITES, the Ramsar Wetlands Convention, the Global Environment Facility, and the World Conservation Union (IUCN). South Africa suggested adding petrels to the agreement, and an option was included to extend the convention eventually to include Northern Hemisphere species as well.

Australia agreed to provide funding and logistical support for another year and pushed for another meeting, which New Zealand and the United Kingdom also agreed to help fund, to conduct the final round of ACAP negotiations. South Africa's minister of Environmental Affairs & Tourism Valli Moosa opened this second meeting in Cape Town in 2000 with participants from sixteen countries, four intergovernmental agencies, and several nongovernmental organizations. Moosa reported alarming effects of pirate longline fishing around the Prince Edward Islands, which South Africa was about to nominate for World Heritage listing, and emphasized the need for the pending agreement to include as a major component efforts to curb illegal fishing. The draft agreement was adopted by consensus at the Cape Town meeting, and the final document opened for signatures in 2001 in Canberra, where the interim secretariat was located.

Australia was the first nation to ratify the agreement, offering to continue as interim Secretariat. New Zealand ratified it a month later, and Ecuador, Spain, and South Africa followed, putting the agreement into force in 2004. Since then, the United Kingdom (2004), Chile (2005), France (2005), Peru (2005), Argentina (2006), Norway (2007), Brazil (2008), and Uruguay (2009), totaling thirteen nations, have ratified and accepted the provisions of the agreement.

Members of eleven albatross and petrel range states or distant-water fishing nations and nine observer and international nongovernmental organizations attended the first gathering in Hobart in 2004. Two South American nations participated, Argentina and Brazil. France, Norway, the United States, and Namibia attended as interested range states and fishers. BirdLife, CCAMLR, SCAR, the

Commission for the Conservation of Southern Bluefin Tuna, and the Convention on Migratory Species also sent representatives. This broad configuration of participation held great promise for what ACAP might be able to accomplish. In many people's minds, ACAP held the potential to enact its acronym and become the capstone of the weblike environmental regime that was coming into place to save the albatross.

At the first meeting, chaired by the director of Australia's Antarctic Division, a fifteen-member Advisory Committee was established. Two working groups were charged with reporting to the advisory committee, the first on status and population trends of albatrosses and petrels and the second on controversial taxonomic splits in three albatross species. Ecuador suggested the Galápagos Islands as the official headquarters, but members decided on Hobart, which is the major center for South Polar and Southern Ocean studies, thus deferring the eventual inclusion of the Northern Hemisphere in ACAP jurisdiction. By placing its Secretariat in Hobart, ACAP chose at the time to have easier access to regional fisheries bodies and researchers focused on sub-Antarctic islands and the albatross latitudes.

The FAO-driven 1995 Code of Conduct for Responsible Fisheries and its 1999 IPOA—Seabirds provide a structure for operational procedures by which fishers and fleets of various nations can minimize marine bird casualties. The challenge for those agreements is the difficulty in persuading parties to respond voluntarily to soft laws that are nonbinding and advisory in character. ACAP is a hard-law treaty, meaning its provisions are binding on its members under international law. It is the culmination of years of effort, especially on the part of Australia and New Zealand, to provide hard laws to save seabirds. But it has limited leverage when a nonparty stakeholder, despite being a member of a fishery management body, the FAO, or some other soft-law association, chooses to ignore or quietly circumvent ACAP regulations and requirements.

ACAP has now moved from a regional to a global agreement that encompasses all land and water areas inhabited by the twenty-two albatross species—including three North Pacific birds—and the seven petrel species it protects. It now includes the island possessions of the United States, Japan, and Mexico in the Northern Hemisphere and islands belonging to France, Ecuador, Chile, South Africa, Australia, New Zealand, and the United Kingdom, plus disputed claims, notably to the Falklands, in the Southern Hemisphere. The agreement's Action Plan recognizes the authority of existing international instruments with conservation measures important for seabirds, including CCAMLR, the tuna fishery management bodies, and FAO's IPOA—Seabirds.[2]

The agreement was tailored to include all nations on a sliding-fee scale. ACAP recognizes the need for capacity building and provides needed support to range states and developing countries. It was envisioned that member-state governments, by working together through ACAP, would be able to integrate their National Plans of Action for Seabirds as well as more forcefully advocate for seabird bycatch mitigation and enforce compliance in the regional fishery management bodies to which they belonged. Parties agreed to conduct research, monitor and restore bird populations, manage fishing, reduce pollution of the sea, and in general support the actions detailed in the International Plan of Action for Seabirds. The Secretariat serves as a depository for national reports that could be used to build a specialized database for ACAP planning and decision making. The poor history of nations' timely submission of standardized information and the uncertainty of ACAP's ability to construct a useful database that is maintained and updated on a timely basis remain obstacles to progress.

Members also agreed to require their fishers to adopt the bycatch mitigation measures of the fishery management bodies to which they belonged and to combat illegal fishing. On the basis of the precautionary principle, they agreed to limit total allowable catches when fishers reduced or harmed the status of seabird species or when they failed to follow FAO guidelines for reducing incidental losses on longlines. The parties adopted a promising set of dispute resolution procedures. The Law of the Sea and its 2003 provisions that deal with migratory fish stocks and compliance by fishing vessels had established an international body to hear marine issues, including fisheries disputes. But experience with Law of the Sea dispute resolution over tuna fisheries does not bode well. A conflict between Australia and Japan over harvest quotas for Bluefin Tuna went into arbitration, but on a technicality the matter was sent back to the parties. The court ultimately failed to decide or rule, which effectively nullified the value of having an international appeals body. Under the ACAP dispute resolution agreement, however, after conferring with the disputing parties, the chair of the advisory committee is to foster an amicable resolution of the dispute. If there is no resolution after one year, the chair establishes a technical arbitration panel made up of advisory committee members and other appropriate experts to resolve the dispute. Decisions of the arbitration panel are binding, which gives ACAP extra backbone. Any dispute about the interpretation or implementation of ACAP provisions will be referred to the body with jurisdiction over the area of concern, such as CCAMLR. Parties will reconcile differences by adopting the policies of the applicable fishery management body for reducing the incidental take of albatrosses. If it wishes, a party or parties may implement stricter measures to protect the seabirds within their sovereignties. ACAP does not spell out sanctions, selective incentives, or degrees of obligation for signatory states, and is unlikely to add any, lest it undermine the entire convention by spreading suspicion and mistrust among members.

The Advisory Committee serves as the hub for ACAP plans and operations. At its first meeting, the committee decided that ACAP would seek observer status at future meetings of fishery management bodies, specifically meetings of CCAMLR and of the four tuna fisheries commissions in the Atlantic, Indian, and Pacific basins, and at future meetings of the FAO's Committee on Fisheries. This has been accomplished. ACAP members can now discuss firsthand with these bodies the problems arising from the overlap between foraging ranges and commercial fisheries and help develop mitigation approaches best suited to those areas. A working group was set up to build a database for breeding colonies to identify areas for eradication of nonnative species that threatened nest sites.

Nine parties and Brazil attended the second full ACAP meeting in Christchurch, New Zealand, in 2006. By then the Secretariat had established observer status with regional fishery management bodies as agreed upon previously at the Advisory Committee meeting. No change in the status of any bird listed in the Annex had occurred, although adopting some of the recommendations of the working group on taxonomy, the parties agreed to remove Gibson's Albatross and the Pacific Albatross from full species standing and confirmed a separation between the Shy Albatross and the White-capped Albatross.[3]

ACAP adopts a precautionary approach in its conservation measures, affirming that it is not necessary for decision-makers to have scientific certainty in order to prescribe measures on behalf of albatrosses and petrels. Establishing the certainty and extent of harm being done to a population or quantifying all the likely effects of human activity in a species' range likely will never be possible in complex ecosystems with all the natural and human variables that govern them. Use of the precautionary perspective allows ACAP decision-makers to broaden their approach to wild animal management from one based solely on empirical evidence—which in itself can be inadequate, skewed, flawed, or otherwise problematic—toward a more contextual one, relying on members to reach decisions using the best knowledge available at the time. It is by design that a broad range of stakeholders and interested parties articulate various positions, assess risks differently, and suggest different strategies for action. This approach opens up a valuable dialogue about environmental issues, including the protection of seabirds, on a wider, more inclusive basis—a dialogue that is itself part of the solution. On the downside, all this discussion can just provide a shield behind which decision-makers avoid dealing with environmental problems that are too politically or economically fraught.

Political scientists have evaluated ACAP and its effectiveness. At the time ACAP was formed, a comprehensive and overlapping network of international principles, standards, laws, and agreements had been slowly developing and coming into force. That emergent structure, full of disparate energies and forces, was in place as a supporting framework for a hard-law convention such as ACAP to draw on and surpass. But ACAP hasn't fulfilled its promise, and many of its proponents are disheartened.

The chief players in the Southern Hemisphere are on board, and now that Brazil has come in, the set of range states is complete. Australia continues to play a leading role, but in general, governments have not engaged in or committed themselves to the ACAP program. The agreement's costs to date have exceeded its benefits, but perhaps that was unavoidable for a start-up agreement. But after six years of being in force, it is not unrealistic to expect that ACAP's benefits begin exceeding or at least balancing costs. Some member countries have sent lawyers as their representatives instead of scientists or fishers or government and nongovernmental officials skilled in negotiating and in developing effective strategic plans. As a result, things have bogged down. ACAP has failed so far to interest Northern Hemisphere governments. France and Spain have not attended a meeting, and Japan never came to the table in the first place. The United States is not a party, although it is currently considering signing despite earlier misgivings about the effects of membership on offshore drilling.

Many people are disappointed in ACAP's failure to add value or force to the preexisting yet incomplete structure for albatross conservation. It appears that ACAP has not yet been able imaginatively or strategically to take the position of leadership, authority, and support its framers designed for it. Only by figuring out how to move into that capstone position can ACAP integrate, reinforce, and extend the efforts already being made to reduce bycatch by nations, regional fishery management bodies, fishers and trade associations, and nongovernmental organizations.

We've gone from Robert Cushman Murphy and his bride, Grace Emeline, to an international Agreement for the Conservation of Albatrosses and Petrels (ACAP) that has yet to come into its place of prominence and leadership in the environmental regime for fish and seabirds from which it emerged. ACAP badly needs full engagement of its national members and highly skilled, forceful leadership in extending enforcement of its hard-law structure throughout the Southern Hemisphere. It's not too late, and the story is not over. ACAP brings us to a potential-filled resting point in this story of albatrosses and fish. By now we hope that it is more than clear, that it is also highly motivating, to see how the expressions and actions of individual bird enthusiasts, bird scientists, environmentalists, governments and government agencies, nongovernmental organizations, artisan and commercial fishers, private organizations, international conventions and laws, princes, sailing heroes, and other stars have come together in what we now recognize as the structure and materials of a still incomplete environmental regime for the albatross.

Hope

CAN ALBATROSSES SURVIVE our invasion and sophisticated exploitation of their ocean habitat? For some populations and even some species, probably not. Not because many fishers intend to kill albatrosses, but because albatrosses are their "incidental catch." But with hundreds of millions of hooks cast into the world's oceans annually, and a chronic background of mortality associated with trawl fisheries, this bycatch is not so incidental. Some experts believe the figure of three hundred thousand seabird deaths from fishing each year, an extrapolation from reports by the relatively small number of onboard observers, to be too low by a third. Add to that the significant numbers of albatross eggs, chicks, and adults also being killed in their nest colonies by introduced predators on islands where habitat has not yet been restored from sealing, whaling, and settlement. K-selected species like albatrosses, which hatch a single chick every one or two years or so, cannot sustain these mortality rates.

Does it matter that tens of thousands of albatrosses are dying at sea every year and that theirs is the most endangered bird family in the world? The answer depends quite literally on the way we see ourselves in the world, as part of an interlocking whole, or as individuals with more local and immediate concerns seemingly unrelated to the whole. The albatross is not a commonly observed seabird, and most of us will never see one. It has no directly significant economic value, although ecotourism boosts its imputed economic value. Yet against all odds, significant efforts and resources are now being expended on saving these birds from extinction because many people, including artisanal and commercial fishers, now regard them as an integral part of a larger whole. Albatrosses, like their cousin petrels and shearwaters, are superbly adapted to ocean ecosystems, but it is not known what effects their diminishment and disappearance would have on these systems. We are mining these marine areas of native fish while increasingly recognizing that our sources of livelihood are indeed part of intricately balanced ocean ecosystems. At the same time, a worldwide web of committed individuals, organizations, and governments is working with commercial fishers to reduce bycatch deaths.

We've come a long way. The puzzle we began the story with—Robert Cushman Murphy's voyage to find out what an albatross was—evolved into that of connecting the dots to discover the vast circumpolar home and range of the albatross. Along the way, it became clear that the albatross had paid a dear price for its interactions with sealers, whalers, and farmers. Then the picture became a puzzle again when commercial high-seas high-tech fishing became possible. It was a few scientists, working in relative isolation on remote Southern Ocean island nest sites, who discovered the connection between fishing and albatrosses—which revealed a more complicated puzzle, that of saving fish as well as the albatross and other endangered seabirds. At this point, the puzzle became three-dimensional, something we had to begin describing in terms of foundational components and structural materials. What can truly be called an international environmental regime to save the albatross has emerged and appears to be solidly established, albeit suffering from some foundational weaknesses and missing struts and beams.

We've seen the results of those weaknesses. Despite the strength of foundational international agreements and laws in regard to fish and fishing, migratory and endangered species, and seabird bycatch, it is a fundamental weakness that all of these agreements depend on the will of individual nations for their enforcement. A second foundational weakness is the lack of adequate data for sound decision making in regard to fish and fishing, seabirds, and marine ecosystems. Fishery management bodies, individual nations, and individual consumers are all critical structural components in the environmental regime to save fish and seabirds. But the structural materials they provide are greatly stressed and weakened by lack of effective leadership and national political will. Fishery management bodies that are trying to manage their fish stocks and address the problem of incidental catch are severely thwarted by inadequate data and lack of enforcement by member nations. Their management decisions are for the most part still focused on single species and suffer from the difficulty of developing accurate population models for those species. Management bodies like CCAMLR and WCPFC, which are committed to ecosystem management, face the seemingly impossible challenge of modeling entire marine ecosystems. Nations sign fishing codes and bycatch understandings but choose to ignore them when it comes to their own fishing industries. Nations provide fishing boats with flags of convenience but impose few regulations on them. Nations turn a blind eye to their own illegal fishers and allow illegal catches of diminished-stock fish to be landed in their ports. Some nations fail even to come to the table to deal with endangered fish stocks and bycatch, as we have seen with ACAP. Some developed nations exploit fish stocks of developing nations that have not yet been able or willing to manage their own exclusive economic fishing zones for their own national interest. Consumers, who may lament the endangered status of fish and seabirds, fail to see the connection between their purchasing decisions and the status of these animals. Or they see the connection but think their buying won't make a difference, since everyone else is buying as well, which is the free rider problem in reverse. It is well known who most of these nations are, and they include many of the developed nations in the Northern Hemisphere. It is well known which nations allow illegal fishing and landings of illegal fish, largely Southern Hemisphere and developing nations, but, scandalously, this list includes some of the leading developed nations. It is well known where the major consumption of endangered stocks occurs—in developed nations in the Northern Hemisphere. In addition to these foundational and structural weaknesses, the Agreement for the Conservation of Albatrosses and Petrels, designed to be the capstone of the environmental regime to save the albatross, has not yet taken its place.

There's a role for everyone in this. Developing nations like South Africa and Ecuador in many instances have been leaders and can provide models in conservation while protecting artisanal economic security as well as the sources of fish protein within their waters. Many developed nations, like Australia, the United Kingdom, and New Zealand, have stepped out to take forceful positions and actions to conserve their own marine ecosystems. On the other hand, many of the developed nations have made limited or meaningless commitments to marine conservation because of their greater stakes in protecting things as they are in the global market. As we're completing this in late 2010, it has become dramatically clear that "things as they are in the global market" is now things as they were.

We did not know, when we began this book, how the story would end and whether we would be in a position of hope or despair. We were determined to find hope if we could and build on that. And we did, describing the sea change that appeared to be happening in regard to fishing and seabird conservation. But we could not escape the possibility that, as with whales, all this effort was too late. What we did not expect was the 2008 global financial crisis that is bringing about a global change in the financial market and economy and their relationship to—and dependence on—sound governance. With the collapse of liquidity worldwide has come an economic crisis that at this point economists do not entirely understand. Nor are they able to predict from day to day the consequences for the financial system, the economy, or civil society. Many experts not only argue that this turn of events was inevitable but also believe it provides us yet another opportunity to try to get it right. And it is highly probable that efforts to get it right will no longer be able to exclude the costs of human productivity and consumption and the borrowing necessary to both. Until the current economic crisis, people have been insulated from, or able to turn a blind eye to, these costs. Capital and labor have been "leveraged" by increasingly sophisticated elaborations of debt instruments. But natural resources and the externalities associated with using them may be beginning to find their place in productivity and profitability calculations. This could provide a harsh and inescapable reality check on the financial elaboration of capital. How do we leverage natural resources? How do we leverage fish or seabird bycatch? Therein lies the paradox, because to get more marine resources, we have to take less. As a result of the current global financial crisis, a fundamental reallocation of resources needs to occur. The human capital that has been seduced into financial engineering may now be drawn, through government incentives and spending, toward research into biotech and energy technologies as well as toward restoring and maintaining critical infrastructures, habitats, and natural living systems. What a hopeful picture that is.

In the meantime, we are killing a mysterious and awesome group of regal birds. They inhabit a domain that frightens us, live in a saltwater world that for us is the alien landscape of vast, monotonous, and uninviting stretches of the earth's surface. They fly enormous distances, sleep both on the wing and on the water, and show an uncanny knack for finding their way. They are large and fearless, seemingly courteous and solicitous with their mates and docile on their nests. They appear to have an unassailable commitment to guarding their single egg or new chick. They manifest what we characterize as a natural wisdom in a life that is, until we intervene, as lengthy as ours.

What disturbs us is the dramatic contrast between the living bird and the transmogrified carcass winched aboard an industrial longliner. We lament the sodden buckle of sinew and bone that was a bird superbly adapted to its watery world a few minutes or hours ago. We bemoan the sodden fling of a long wing as one, then perhaps another, individual comes over the side, followed tomorrow by another one or two from another set, and several more in the days ahead. The fishers cut their limp bodies loose and toss them back into the ocean, regretting lost time and lost fish. For scientists and conservationists, these carcasses are a waste of genetic capital. The trickle of dead birds tumbles quietly and drifts, without drama, into a population loss that can only be discerned by fieldwork in the harshest or most distant of environments. Bycatch losses are compounded by losses from birds that cannot differentiate between fish and plastic or other floating debris and who can't defend themselves from alien invasive predators.

How do we halt the destruction of such superb birds that were planing the ocean swells long before any humans appeared in the archaeological record? We can shrug and turn away. We can rationalize the losses because only the fittest survive anyway, there is a price to be paid for feeding the world, more winners than losers exist in the expanding global economy, and surely this problem is a mere blemish in the history of our ability to master the environment with our miraculous technology. We can dismiss albatrosses as anthropomorphized metaphors of a once clean, fresh world, or merely as objects among others that fell within our purview to be named and studied. And as we do this, the birds are dying; their numbers are dropping, and what we call species are disappearing, maybe for good.

The other approach is that explored by individuals who embrace the world as an open system of links that bind life together in all sorts of manifestations. They recognize concepts of human priority and entitlement as tragically shortsighted, counter to reality and radically self-defeating. There is more at stake in the joy and fascination of seeing the soaring pinions of a sea-spun albatross, the broad shadows of a banyan tree, or the compound eye of a praying mantis. We need them all, or at least we need to limit our activities that take out entire plant and animal species. Many losses have occurred due to ignorance or callousness. But scientists and conservationists have seen to it that we cannot now dismiss the plight of the albatross out of ignorance. We know what is happening, although it has not been easy to put together the explanatory picture of their decline and then fabricate a global regime to save them. The interdependence of the various stakeholders, organizations, and governments is an expression of collaborative governance—governance that does not transcend the priority of the nation but instead engages multiple nations in a shared priority, that of managing resources that future generations have a right to enjoy. In all this, the role of nongovernmental organizations is clear. They provide the leaven. They provide the science and the data, the leadership, the provocation, the interest brokering and negotiations, and then much of the daily on-the-ground-or-water work. Scientists and government officials also play some of these roles and act as leaven. As do individuals. But there is no substitute in this work of collaborative governance for the contribution of environmental organizations with global constituencies. Nor is there any substitute in market governance for the contribution of responsible consumers, who hold in their hands the as-yet-unrealized and -unclaimed power to shape the market for sustainable use.

We have not yet solved the challenge of how to exercise effective control over the high seas and protect the planet's marine resources. In the best scenario, individual scientists will continue their fieldwork and furnish up-to-date information, document declines, and suggest priorities and techniques for addressing the problems faced by plants and animals. Commercial fishers and fisheries bodies will continue to develop ways of managing fish stocks sustainably and fishing without destroying seabirds and sharks and turtles. Celebrities, sports figures, authors, filmmakers, and broadcasters will spread public awareness of the threats to fish stocks and albatross survival. Bird and wildlife groups will continue to expand their leavening role that makes environmental regimes possible. And consumers will empower themselves and take back the market—they hold the most radical potential for change of all these groups. Person by person, government by government, organization by organization, and international agreement by international agreement, we can pray that the seas will continue to be home to their iconic albatrosses, slanting and soaring in the salt and spray as much as in our mind's eye. This has been a narrative of hope.

Appendix

CCAMLR AND SEABIRD MORTALITY

DATE	ACTION/EVENT
1984	US delegates submit a report that addresses the incidental losses of wild animals from discarded fishing gear and "certain fishing operations" (Haward, Bergin, and Hall 1998, 260). CCAMLR requests members to document any birds and marine mammals taken incidentally during fishing operations.
1986	An ecosystem-monitoring program is set up to track changes in the ecosystem, especially predation on krill and consequences for fishing. The Scientific Committee uses these data to update the status and population trends in marine birds and mammals within the area. CCAMLR requests that area fisheries record and report bird losses during commercial operations.
1987	Longline fishing vessels begin to exploit mature toothfish in areas inaccessible to trawls.
1988	Fisheries observers gather seabird bycatch statistics on Japanese longline tuna vessels in Australian waters. Members are allowed to place inspectors on vessels of other contracting parties within the CCAMLR area to ensure they comply with CCAMLR requirements.
1989	The incidental catch of seabirds is made a separate agenda item at both the meetings of CCAMLR parties and CCAMLR's Scientific Committee. CCAMLR adopts Resolution 5/VIII about protecting seabirds from mortality due to longline fishing. The Scientific Committee discusses the potential problem to nesting seabirds, especially as demersal longline fishing has begun within the CCAMLR region (South Georgia statistical subarea 48.3). CCAMLR establishes catch limits for the Patagonian Toothfish longline fishery for the 1990–1991 season and requires reports of seabird entanglements or losses in that fishery.
1990	Australia presents a paper to the Scientific Committee that estimates that pelagic longlines outside the CCAMLR area kill 44,000 albatrosses annually. It describes how collaboration with Japanese tuna vessels in using "bird-scaring

DATE	ACTION/EVENT
	streamer lines" reduces seabird mortality. CCAMLR endorses the 1989 UN driftnet resolution that no expansion of driftnets will occur in CCAMLR's area (Resolution 7/IX).
1991	CCAMLR issues a conservation measure (CM 29/X) requiring action to minimize incidental mortality to seabirds in fishing or research with longlines within its area. CM 29/X provides for quick sinking of lines and night setting; prohibits dumping offal during commercial or research longlining; and specifies streamer type, method of deployment, and use during daylight setting. Entered into force in 1992; revised, updated, and continues as CM 25-02. Real Time Monitoring Program (RTMP) established in Japan's tuna fishery for catch evaluation; it is extended to bycatch. First conservation measure requires toothfish vessels in the CCAMLR area to deploy seabird-scaring devices and instructs observers on vessels to increase note of seabird casualties.
1992	Scheme of International Scientific Observation reinforces vessel inspections.
1993	Scientific Committee establishes an ad hoc Working Group on Incidental Mortality Arising from Longline Fishing (WG-IMALF). Abbreviated to WG-IMAF (after 2001 to include all fishing, including trawls), the group maintains a database on breeding albatrosses and petrels in the CCAMLR Area and an at-sea distribution of the species in relation to FAO Statistical Areas, Subareas, and Divisions. The group reports to the Scientific Committee. CM 29/XII mandates only thawed bait be used on longlines. The Scientific Committee notes major declines in toothfish stocks around South Georgia.
1994	WG-IMALF reports the serious albatross losses within the CCAMLR area and around the boundaries. CM 29/XIII approves of unavoidable dumping only on the side away from line use area. It requests that all efforts be taken to release and remove hooks from live birds. CCAMLR outlaws the use of net sonde cables that kill seabirds on bottom trawls.
1995	WG-IMALF (reporting under WG-FSA; Working Group on Fish Stock Assessment) calls for more accurate data about bird losses and an assessment of devices deployed in reducing seabird deaths. CM 29/XIV specifies weights for the Spanish system of longlining and mandates line setting at least 3 hours before dawn to reduce losses to White-chinned Petrels. Working Group on Fish Stock Assessment reports that the illegal take of toothfish is as high or higher than the reported catch. Demersal (toothfish) longlining is clearly implicated in seabird deaths.
1996	CCAMLR distributes *Fish the Sea, Not the Sky*, an educational text for fishers. Toothfish make up 99% of the reported catch of finfish (among 13 species sought commercially five years earlier) for the entire region. United Kingdom

reports illegal fishing around South Georgia and South Africa's Prince Edward Islands. CM 25-01 regulates the disposal of plastic packaging bands on vessels.

1997	CCAMLR delays the opening of the toothfish fishery over much of the area from March 1 until April 1 (in 1998), then until May 1 (in 1999). CCAMLR discusses IUU fishing: WG-IMALF estimates losses at 5,700 birds (40% albatrosses) in the Atlantic Ocean sector, 1,000 birds (23% albatrosses) in the Indian Ocean sector, with an overall estimate of 66,000–107,000 seabirds lost to IUU fishing. This is twenty times larger than for the regulated fishery. CM 29/XVI prohibits dumping of fish discards during setting and permits only unavoidable dumping during hauling.
1998	CM 10-04 requires (by December 31, 2000) a vessel monitoring system (VMS) for all boats licensed for toothfish in the area. Estimated mortality of seabirds exceeds 7,000 from regulated, but mainly IUU, fishing in the area (notably the Indian Ocean).
1999	Catch documentation scheme (CDS), established to counter IUU fishing, requires the inspection of toothfish vessels licensed by a contracting party when entering the port of another CCAMLR party. Becomes binding on members on May 7, 2000.
2000	CM 29/XIX revises Spanish line weighting to 8.5 kg at no more than 40 m spacing, and permits daylight setting with a minimum sink rate of 3 m/second. It allows fishing only if vessels can process offal or discharge it on the opposite side of line use. Norway proposes a resolution that is adopted urging parties neither to flag nor license nonparty vessels in EEZ waters if they have a history of IUU fishing in the CCAMLR area. It is, however, not a binding conservation measure as Norway had proposed. Experts speculate that IUU has taken up to 68,300 albatrosses from a total of as many as 257,000 seabirds in the last four years in the area.
2001	CM 148/XX implements an earlier measure for a satellite-linked vessel monitoring system on toothfish vessels fishing in CCAMLR waters. Mauritius, a major player in illegally fished toothfish landings, agrees to implement CDS requirements.
2002	An estimated 6,500 seabirds are lost to IUU operators in the area. WG-IMAF calculates area losses from longlines in previous six years at 270,000 to 700,000 seabirds. CM 29, now 25-02, requires vessels to remove hooks from fish heads and offal before discarding.
2003	WG-FSA reports losses of 15 seabirds in the area's regulated longline fishery— the lowest on record. However, the working group estimates that 20 species of albatrosses and petrels are at risk from longlining in the Southern Ocean. CM

DATE	ACTION/EVENT
	25-02 specifies the weight for the Norwegian Autoline system (5 kg at 50–60 m spacing, or an integrated weight of 50g/m). It abolishes the need for frozen baits because current weights provide acceptable sink rates and revises specifications for streamer lines, encouraging two per set. CCAMLR draws up a list of IUU vessels and approves an electronic web-based CDS trial. CM 10-06 sets up a Black List that stops vessels from unloading fish in 34 nations' ports. CM 25-03, on trawling, prohibits net monitoring cables and offal discharge, whenever possible; advocates clean nets before shooting; and limits time nets remain on sea surface.
2008	CM 10-03 requires port inspection of toothfish vessels and use of automated VMS and CDS for toothfish. In addition, a scheme is originated to promote compliance from nonparty vessels in the CCAMLR area. As of 2007, CCAMLR places observers aboard all longline vessels, endeavors to eliminate seabird bycatch by working to improve underwater setting and other devices, works toward similar mitigation efforts while seabirds forage outside the official area, pursues IUU fishers and collaborates with other fisheries bodies and international organizations, and encourages its members to adopt IPOA— Seabirds and formulate national action plans. CM 25-02 as amended requires longlines to sink as rapidly as possible, integrated weights be used for autoline systems, night setting, no offal dumped if possible, removing hooks from fish heads, releasing birds alive, use of streamer line(s) on line setting, and device(s) to reduce bait access in hauling. CM 24-02 Longline Weighting for Seabird Conservation sets out various protocols to be selected, implemented, and monitored by onboard fishery observers.

Sources: http://www.ccamlr.org/pu/e/gen-intro.htm; Blue Ocean Institute, blueocean.org; Constable 2001; Tickell 2000; Haward, Bergin, and Hall 1998; CCAMLR 2007, 2009b.

Notes

CHAPTER 1

1. See BirdLife International 2010c, "Species at Risk," under Save the Albatross Campaign.
2. Murphy 1965, 148 (quote).
3. These are lines 115–118 in Coleridge's 1798 *Rime* cited by Martin Gardner (1965, 121). Gardner notes that Coleridge added seven lines and removed forty-six in the second edition of *Lyrical Ballads* in 1800, and made additional changes in another version in *Sibylline Leaves,* published in 1817.
4. Coleridge's 1798 *Rime,* lines 611–618 and 645–650, cited by Gardner (1965, 160 and 165).

CHAPTER 2

1. Bell and Robertson 1994, 219–228; Robertson 1991, 5, 7, 54–55, 62–63, 68–70. Robertson mentions the 1921 Animal Protection and Game Act as protecting Royal and Buller's Albatrosses, and the regulations gazetted on March 17, 1931, added the Chatham Albatross to the protected schedule (1991, 9).
2. Banks, "Banks's Journal," February 1769; Tickell 2000, 358. John Dunmore (2006) envisions the captain's wife, Elizabeth Cook, testing recipes, including stewed albatross, based on input from many mariners she had met, and shows an albatross on the cover of the cookbook.
3. Williams and Simison 1944, 42 (quotes).
4. Medway 1998, 189–198, 189 (quote).
5. Tickell 2000, 358. Murphy (1965, 243) reports that Captain Cleveland sold albatross and penguin eggs to curio dealers in New Bedford and Cape Cod for "a dollar apiece" (quote).
6. The International Union for the Conservation of Nature and Natural Resources was founded in 1948. Use of the name World Conservation Union began in 1990, but the original acronym is often used because people still know the union as IUCN.

7. Matthews 1931, 83 (quote).

8. Doughty 1975.

9. Rauzon 2001, 107 (quote); Tickell 2000, 198, 211–214, 243.

10. M. L. Fisher 1970.

CHAPTER 3

1. Tickell 2000, 14, 357, referring quotes to Markham 1878. For Shelvocke, whose 1726 *Voyage Round the World by the Way of the Great South Sea* includes the albatross that William Wordsworth, then Coleridge's most trusted friend, suggested be featured in a work of Romantic realism that Coleridge intended to write, see Gardner 1965, 46–47.

2. Murphy 1936, 1:540 (all quotes).

3. Ibid., 1:540 (quote).

4. Murphy 1965, 149 (quotes).

5. Murphy 1936, 1:220 (quote).

6. Ibid., 1:570 (all quotes).

7. Ibid., 1:556–557 (all quotes), and repeated in *Logbook*, 243.

8. Murphy 1965, 160 (quote).

9. Ibid., 165 (all quotes).

10. Ibid., 173 (quote) and 175.

11. Murphy 1936, 1:556 (quote); Headland 1984, 119.

12. Murphy 1965, 194 (quote).

13. Ibid., 194 (quote).

14. Ibid., 200 (quote).

15. Ibid., 200 (quote).

16. Matthews 1929, 563, 568.

17. Matthews 1951, 13 (quote).

18. Ibid., 13 (quote).

19. Ibid., 125 (quote).

20. Adrian J. Howkins (2008, 17) notes that the Scott Polar Research Institute was established in Cambridge, England, in 1920 due to the Discovery Expedition surveys that fostered enduring links between science and politics in Antarctica.

21. Ian Wallace (2004, 73) assigns this quote about WB to William (Bill) Bourne. Information about the Edward Grey Institute is also derived from Tim Birkhead (2008, 349), who quotes William H. Thorpe (1974, 271–293).

22. C. J. R. Robertson 1998b, 4:430 (quote).

23. Richdale 1950, 7 (quote).

24. Ibid. (quote). Richdale began a thirteen-season study from November 1936 to May 1949, consisting of 484 visits to the Taiaroa Head colony.

25. Fleming 1984, 27–37; Richdale 1939, 467–488; 1942, 169–184; C. J. R. Robertson 1993b, 269–276; Tickell 2000, 55.

26. Robertson and Wright 1973, 49–58; Robertson 2001, 7–8.

27. Southern Seabird Solutions 2007 reports that 170,000 visitors arrive on the headland every year to view the colony. Perriman notes that careful and calm handling assists young birds in becoming less fearful and aggressive toward human managers. He notes calm parents usually produce calm offspring; personal communication, August 11, 2010.
28. Richdale 1952, 83 (quote); C. J. R. Robertson 1998b, 430.
29. Waring and Robbins 1996, 928–930.
30. Safina 2002, 271 (quote).
31. Fisher 1966b, 221 (quote).
32. Tickell and Morton 1977, 363 (quotes); Tickell 1975, 125–131; 2000, 240–243.
33. Hannon 2006; Tickell 2000, 239, 245; Steiner 1998; US Fish and Wildlife Service 2005, 2, 4; BirdLife International 2008e.
34. See Steiner 1998.
35. For information on the transfers, see Deguchi 2010.

CHAPTER 4
1. C. J. R. Robertson (1993b) notes that Grandma wore several bands throughout her breeding life, and was wearing a stainless steel R15113 and an added combination of blue/white when she disappeared. Button raised a chick in 2010, and Perriman reports that he has "calmed down" since being handled and fed as a youngster, which reportedly added to his aggression; personal communication, August 11, 2010.
2. Tickell (2000, 7–9) offers a remarkably detailed recognition of the people who have assisted in making the author's commitment to his birds so productive.
3. Gibson and Sefton 1959, 73 (quote).
4. Ibid., 73–82.
5. Ibid., 78 (quote).
6. Ibid., 73–82, table p. 79.
7. Tickell and Gibson 1968, 7–20.
8. Headland 1984, 102; Tickell 1970, 84–93.
9. Gibson 1967, 47–51.
10. Tickell 1970, 88 (quote).
11. Lindsey Smith, personal communication, SOSSA Pelagic, Wollongong, April 28, 2007.

CHAPTER 5
1. CCAMLR, "General Introduction, 2007." The area covered is spelled out by the original documents to CCAMLR. "This Convention applies to the Antarctic marine living resources of the area south of 60° South latitude and to the Antarctic marine living resources of the area between that latitude and the Antarctic Convergence which form part of the Antarctic marine ecosystem," http://www.ccamlr.org/.

CHAPTER 6

1. Alexander 1954, vi (quote).
2. Croxall, Evans, and Schreiber 1984, based on the proceedings of the ICBP Seabird Conservation Symposium in Cambridge, August 1982.
3. Ibid.
4. Ibid., 2; Gaston 2004, 5–11, 45.
5. See BirdLife International 2004b, 25.
6. There has been a discrepancy for Midway, where 286,000 nests in 2001 had risen to 441,000 in the 2003 census, raising a concern about accuracy in counts. Recent data are in Naughton, Romano, and Zimmerman 2007.
7. Liittschwager and Middleton 2005, 86.
8. See Tickell 2000, 16–17, for a discussion of its name.
9. BirdLife International 2004b, 24–25.
10. Mlodinow 1999.
11. See Carpenter 2004; Salkeld 2007; and ACAP 2010a.

CHAPTER 7

1. Doughty 1975, 97 (quote).
2. For a startling account of current exploitation, read Franzen 2010.
3. Lyster (1985, 97) quotes the Western Hemisphere convention's objectives.
4. Environmental Treaties and Resource Indicators (ENTRI) n.d. (quote); Lyster 1985, 64–66.
5. CMS 2004, 19 (quote); CBD 2010.
6. There are fourteen Memoranda species and groups of animals.
7. Burhenne-Guilmin 2004, 19 (quote).

CHAPTER 8

1. Macquarie Island World Heritage Area n.d.; Walton and Bonner 1985, 11.
2. TAAF also consists of Terre Adelie land in Antarctica and a fifth district, Iles Esparses; including the "albatross islands," these cover an exclusive economic zone (EEZ) of one million square miles; see TAAF 2007 and Tickell 2000, 48–52.
3. Hansom and Gordon 1998, 183–184 (quote); Howkins 2008 247–259.
4. SCAR home page lists publications, events, and information pertaining to Antarctica; http://www.scar.org/. Also see Walker 2009, which notes that successful albatrosses must complete their nesting in 140 days, less than five months.
5. Fitzmaurice 2004, 191 (quote).
6. World Heritage n.d. (quote).
7. The World Heritage List is constantly updated. These statistics represent sites in 2010; see http://whc.unesco.org/en/list/577.
8. Vidal 2008 (quote).
9. See Parkes 2008, 5 (quote). For the Heritage List, consult http://whc.unesco.org/en/list/877, and the Marine Programme, http://whc.unesco.org/en/activities/17/.

10. Clover 2004, 339 (quote); CITES at http://www.cites.org/eng/disc/what.shtml, and regarding the Bluefin Tuna, see http://www.iuncredlist.org/search/details .php/21858/summ, which defines this fish's status as reduced by a minimum of 80 percent over a three-generation time span. See Greenberg's "Tuna's End" (2010b), which is included in his new book, *Four Fish: The Future of the Last Wild Food* (2010a).

11. Willock 2002, 34 (quote).

12. Sovacool and Siman-Sovacool 2008, 30 (quote).

CHAPTER 9

1. "Exploit," used in regard to harvesting of living resources, was at first a neutral descriptive term meaning to turn (a natural resource) to one's account, to work, to cultivate. Its meaning has now become so identified with the secondary meaning, to make use of meanly or unjustly for one's own advantage or profit, that it is almost impossible to use it in its neutral sense, but the neutral, commercial sense of "exploit" is the usage we employ in this book. When we need a descriptor of negative effects, we use "deplete," "collapse," "threaten," and "endanger," among other words for which the negative connotation is unambiguous.

2. "Fishery" refers to an area of a major, harvestable concentration of a target stock, such as the now collapsed Black Cod around Nova Scotia, or the Atlantic Bluefin Tuna in the Atlantic Ocean and Mediterranean Sea.

3. Clover 2004, 5 (quote).

4. UN, FAO 2005, 2–4; Clover 2004, 137–140, deals with subsidies.

5. Watson and Pauly 2001, 534–536.

6. Myers and Worm 2003, 280–283, 281 (quote).

7. Most recently, a two-year study by an international team of twenty-one coauthors, led by Boris Worm and Ray Hilborn, showed that steps taken to curb overfishing were beginning to succeed in five of the ten large, highly managed marine ecosystems that they examined. However, of all the fish stocks that were examined, 63 percent remained below target and still needed to be rebuilt (Worm and Hilborn et al. 2009).

8. Upton and Vitalis 2003, 1 (quote).

9. Clover 2004, 156–157; Gianni and Simpson 2005.

10. Lack and Sant 2001.

11. Hardin 1968, 1243–1248.

CHAPTER 10

1. Kock 1994, 3–22; Fallon and Kriwoken 2004, 221–266. In 2010, a second and controversial toothfish fishery underwent certification with the Marine Stewardship Council for the Ross Sea.

2. National Environmental Trust 2004. Seabird bycatch statistics are in Lighthouse Foundation n.d.

3. National Environmental Trust 2004; Lack and Sant 2001; Kock 1994, 6.

4. Trevor Corson 2007, 249; and "Giant Tuna Fetches." This 2010 article noted that this price was similar to a record amount paid for a 440-pound fish in 2001.

5. McCurry 2007 (quote).

CHAPTER 11

1. The United Kingdom proclaimed a two-hundred-mile maritime zone around S. Georgia and the South Sandwich Islands in May 1993, and extended the same limit to the Falkland Islands Outer Conservation Zone in August 1994.

2. Pauly 1995, 430 (quote).

3. Clover 2004, 103–107; on depensation, see Liermann 1999.

4. Myers and Worm 2003, 280–283.

5. Clover 2004, 234.

CHAPTER 12

1. Jouventin and Weimerskirch 1990, 746–749.

2. Weimerskirch 2004; Nevitt, Losekoot, and Weimerskirch 2008; Pinaud and Weimerskirch 2002.

3. Jouventin, Stahl, Weimerskirch, and Mougin 1984, 609–625, 612 (quote).

4. Ibid., 623 (quote).

5. Jouventin and Weimerskirch 1991.

6. Jouventin and Roux (1983, 181) based the new species' separation from the Wanderer on weight, eye, bill, head, and breeding coloration, plus a later start to the breeding season.

7. See "The Waterbird Society" n.d.; Croxall 2004a, 549–550; and Croxall 2004b, B7.

8. Croxall 1979.

9. Croxall 1984.

10. Brothers 1991.

11. Brothers and Foster 1997; Kwok, Yau, and Ni 2002, 377.

12. Croxall 1990a, 1.

13. Ibid. (quote).

14. Dalziell and Poorter 1993, 143–145.

15. Gales, *Co-Operative Mechanisms for the Conservation of Albatross*, 2 (quote); Brothers and Gales 1996, 3; CMS n.d.

CHAPTER 13

1. Brothers, Cooper, and Lokkeborg 1999, 39; for BirdLife's classifications for the albatross, see http://www.rspb.org.uk/supporting/campaigns/albatross/about /species/index.asp.

2. The Inter-American Tropical Tuna Commission (IATTC), established in 1950, runs a Tuna-Dolphin Program for fishers in the Eastern Pacific Ocean that promotes fishing methods and equipment to reduce dolphin deaths, notably in

purse seining. There is also an International Dolphin Conservation Program for which the IATTC provides the secretariat and which came into force as a legally binding agreement in 1999; see M. Elliott and Partners 2003, 1, 62–63; and Hall, Alverson, and Metuzals 2000.

3. Bialik 2009.
4. Kock 1992; Bjordal and Lokkeborg 1996, 31–37.
5. Pohl 2003.
6. BirdLife International 2007b (discusses Sullivan's important work), 2008f.
7. BirdLife International 2008a, 37–38.

CHAPTER 14

1. The parties of UNCLOS stood at 160 in March 2010.
2. Olav S. Stokke, "Effectiveness of CCAMLR," in Stokke and Vidas 1996, 122 (quote).
3. Young 1994, 20–24, 66–74; G. Robertson 2002.

CHAPTER 15

1. Brothers 1991, 266 (quote).
2. Brothers, Cooper, and Lokkeborg 1999, 71.
3. Ibid., 71; also, in Australia, Department of Agriculture, Fisheries and Forestry 2003, 11, it is noted that Japanese vessels agreed to carry observers within the AFZ, who gathered quantitative data on bycatch only after 1988. The Japanese fleet fishes south of the AFZ in adjacent water but has not returned to fish under permit.
4. Brothers, Cooper, and Lokkeborg 1999, 6, 84; Australia, Environment Australia 1998, 25.
5. Australia, Department of Agriculture, Fisheries and Forestry 2003, 84.
6. Southern Seabird Solutions 2003 (quote).
7. Gilman, Boggs, and Brothers 2003.
8. World Wildlife Fund 2009; G. Robertson 2007, 5.
9. Southern Seabird Solutions 2004; BirdLife International, Press release, April 24.
10. Brothers, Cooper, and Lokkeborg 1999, 79–81; Double 2007, 8; Pierre 2007, 6; G. Robertson 2006.
11. Croxall, Interview.
12. Brothers 1996a, 2 (quote); 1996b.
13. Australia, Australian Nature Conservation Agency 1996, 4, 14.

CHAPTER 16

1. G. Robertson and Gales 1998b, vii (quote).
2. UNCLOS, http://www.un.org/Depts/los/index.htm.

3. UNCLOS, Part VII, High Seas, Section 1, (quote).

4. Upton and Vitalis 2003, 4 (quote).

5. UN, FAO 1993.

6. Upton and Vitalis 2003, 5 (quote).

7. Ibid., 5–6 (quote).

8. UN, FAO 1995, 2004.

9. Vidal 2006 (quote).

10. Ibid. (quote).

11. Ibid. (quote); Burgmans 2008.

12. Vidal 2006 (quote).

13. Clover 2004, 286–287.

14. MSC n.d.b, 2010b; Vidal 2006.

15. MSC 2004, 2009; COLTO 2007.

16. Clover 2004, 292 (quote).

17. Vidal 2006.

18. Clover 2004, 282, 284, 296; MSC 2003, 2006, n.d.a.

19. MSC 2007 (quote).

20. Lawrence 2006.

21. MSC n.d.c; Vidal 2006.

22. Forest & Bird 2007b (quote).

23. National Environmental Trust 2004; Lack and Sant 2001; Kock 1994, 6; Jolly 2010; MSC 2010a.

24. MSC 2002 (quote); see also Roheim 2003.

25. Brothers, Cooper, and Lokkeborg 1999, 32.

26. UN, FAO 2010a (quote).

27. UN, FAO 2008.

28. BirdLife International 2008h.

29. John Croxall, personal communication, Cambridge, May 29, 2008.

CHAPTER 17

1. Winegrad and Wallace 2005; Lack and Sant 2001.

2. Croxall 2003, 10 (quote).

3. Ibid. (quote).

4. Traffic 2008.

5. Lack and Sant 2001; Traffic 2008; National Environmental Trust 2004. The National Environmental Trust nonprofit organization, founded in 1994, urges members and the public to "Take a Pass on Chilean Sea Bass," due to IUU fishing.

6. Greenpeace International n.d. CITES Appendix II regulates the trade in 22,000 plant and 4,000 animal species through a permit system.

7. See CITES 2002b, Resolution Conference CoP12 Doc. 4, and CoP13 Doc. 36 at www.cites.org.

8. Fishers Forum 2006, 5.

9. CCAMLR 2007, Table 1.

10. CCAMLR 2005; M. Lack 2008.

11. UN, FAO, Fisheries and Aquaculture Department 2005.

12. WCPFC 2006a (quote).

13. WCPFC 2007.

14. WCPFC 2006b, 15; ACAP 2010c.

15. Small 2005, 37–38.

16. CCSBT 2005, Appendix 2; also Working Group on Ecologically Related Species.

17. IOTC 2007, 3; BirdLife International 2008b (quote).

18. ICCAT 2007a (quote).

19. ICCAT 2007b 3–7; International Seafood Sustainability Foundation 2009.

CHAPTER 18

1. See Australian Antarctic Division, http://www.aad.gov.au/, for program details, and Herr and Davis 1996, 331–360.

2. Dutter 2006; Joyner 1998, 149–154.

3. O'Byrne 2010 (quote).

4. See *Viarsa 1* 2003; and Australia, Customs Service 2003.

5. Australia, Ministry of Fisheries, Forestry and Conservation 2005; Blakely 2009; Oceana 2008.

6. Fickling 2003.

7. "Illegal Fishing in the Southern Ocean" 2003; Hutchison 2004.

8. Senator Robert Hill 1997 (quotes); Oceana, Press release, April 21 1997.

9. New Zealand, Department of Conservation 2009 lists eight albatrosses and six petrels in Appendix Two.

10. Australia, Department of Environment and Heritage 2001b, 86.

11. Australia, CSIRO 2007.

12. UN, FAO, Country Profiles: New Zealand.

13. According to the UN, FAO, "Country Profiles: New Zealand—Fishery Sector."

14. Forest & Bird 2007a (quote).

15. Ibid. (quote).

16. Forest & Bird 2007c.

17. Forest & Bird 2010 (quote).

18. Childerhouse et al. 2003; for Auckland Islands, of which Enderby is one, see http://www.doc.govt.nz/templates/page.aspx?id=33842, and the Global Invasive Species Database, http://www.issg.org/database/welcome/.

19. Reports available at http://www.fishersforum.net.

20. C. J. R. Robertson et al. 2003, 7; Forest & Bird 2007c.

21. New Zealand, Ministry of Fisheries and Department of Conservation 2004. This NPOA is based on a UN prototype published in 1999; see UN-IPOA section 17.

22. TerraNature Trust 2003.

23. New Zealand, Oceans Policy Secretariat 2003, 6–9, 7 (quote).

24. "New Zealand Recalls Squid Boats" 2005 (quotes).

25. New Zealand, Ministry of Fisheries 2008 (quote).

26. See High Seas Task Force, "Halt IUU Fishing," http://www.illegal-fishing.info /item_single.php?item=document&item_id=118&approach_id=16.

27. Falklands Conservation 2004, Table 6.1, p. 13.

28. Sullivan, Reid, and Bugoni 2006.

29. Brown et al. 2001; Poncet 2006; Invasive Species in the UK Overseas Territories 2007.

30. See ASOC 2007, 2–3; Croxall 2008b.

31. Australia, Department of Foreign Affairs and Trade 2005.

32. Useful and illuminating blogs by ATF teams show how local circumstances and personalities direct exchanges and outcomes; see http://www.rspb.org.uk /supporting/campaigns/albatross/.

33. Japan, Fisheries Agency 2001, 1 (quote).

34. Ibid., 2 (quote).

35. Ibid., 3 (quote).

36. Ibid., 6 (quote).

37. UN, FAO, Committee on Fisheries 2009, 7–9.

CHAPTER 19

1. American Bird Conservancy 2002; Winegrad and Wallace 2005.

2. RSPB 2000, 35–36. BirdLife held its own World Bird Festival in Barcelona, Spain, in June 2001, which was attended by three hundred thousand people from 88 countries assisted by 74 BL partners; see BirdLife 2001.

3. BirdLife International 2008c.

4. BirdLife International 2005.

5. BirdLife International 2007c, 3; BirdLife International, Albatross Task Force, http://www.rspb.org.uk/supporting/campaigns/albatross/problem/atf.asp.

6. BirdLife International 2008d, 2002d.

7. Lokkeborg and Thiele 2004.

8. Southern Seabird Solutions 2008; New Zealand Department of Conservation n.d.

9. Southern Seabird Solutions 2006; Southern Seabird Solutions 2005 and "Looking Ahead," *Newsletter*, June 2004.

10. Seafood Industry Council 2006.

11. Southern Seabird Solutions 2010 (quote); Shelly Farr-Biswell 2009.

CHAPTER 20

1. COLTO (Coalition of Toothfish Operators), 2007, "Home Page" (quote), http:// www.colto.org/.

2. COLTO 2009.

3. The Organization for the Promotion of Responsible Tuna Fisheries 2000.
4. OECD, "About the OECD."
5. Marine Resources Assessment Group 2005a; High Seas Task Force 2006, 1 (quote).
6. BBC News 2009 cites Helen Bours (quote).
7. Ibid. (quote).
8. International Tuna Fishers Conference on Responsible Fisheries and Third International Fishers Forum 2005, Yokohama Declaration (quote).
9. Fourth International Fishers Forum 2007.
10. International Association of Antarctic Tour Operators 2006, 1–6; 2007; Bird-Life International 2002b, 2007a.

CHAPTER 21

1. BirdLife International 2002a (quote).
2. BirdLife International 2006c (quote).
3. BirdLife International 2004a; V. Elliott 2004; Marshall 2004 (quote).
4. Nevard 2005 (quote).
5. BirdLife International 2002c (quote).
6. United Nations, Environmental Program 2002.
7. Prince of Wales 2004b; BBC News 2005 (quote).
8. Prince of Wales 2004a (quote).
9. Prince of Wales 2005a (quote); BirdLife International 2006a.
10. Prince of Wales 2005a (quote).
11. Ridgway 2003–2004b (quote).
12. BBC News 2005 (quote).
13. RSPB 2008 (quotes); BirdLife International 2006b.

CHAPTER 22

1. Australia, Environment Australia 2000.
2. ACAP n.d., par. 3.2.1.
3. ACAP 2006b.

Bibliography

ACAP (Agreement for the Conservation of Albatrosses and Petrels). 2006a. Meeting of Parties, doc. 5, revised, "List of Participants." http://www.acap.aq/en/index .php?option=com_docman&task=cat_view&gid=21&Itemid=33 (accessed March 2007).

——. 2006b. Meeting of Parties, doc. 12, "Report of the Advisory Committee," and doc. 30, "Amendment to Annex 1." http://www.acap.aq/en/index.php?option=com_ docman&task=cat_view&gid=19&Itemid=33 (accessed March 2008).

——. 2010a. "An Atlantic Yellow-nosed Albatross Visits the United Kingdom." Newsflash, Aug. 14, http://www.acap.aq/latest-news/an-atlantic-yellow-nosed-albatross-visits-the-united-kingdom (accessed December 2010).

——. 2010b. "Species Assessment." http://www.acap.aq/acap-species (accessed October 2010).

——. 2010c. "Western and Central Pacific Fisheries Commission to Consider How to Reduce Seabird Mortality." Newsflash, Aug. 6, http://www.acap.aq/ (accessed August 12, 2010).

——. n.d. List of Parties, Action Plan, Species Assessments, News, etc. http://www .acap.aq/acap-species (accessed 2004 and onward).

Adger, W. Neil, Tor A. Benjaminsen, Katrina Brown, and Hanne Svarstad. 2001. "Advancing a Political Ecology of Global Environmental Discourses." *Development and Change* 32: 681–715.

Agius, Emmanuel, and Salvino Busuttil, eds. 1998. *Future Generations and International Law.* London: Earthscan.

"Albatross." 1850. *Littell's Living Age* 24: 264–265. Littell_s_living_age-pdf (accessed July 2010).

Alexander, Wilfred B. 1954. *Birds of the Ocean: A Handbook for Voyagers.* New York: Putnam.

American Bird Conservancy (ABC). 2002. *Sudden Death on the High Seas. Longline Fishing: A Global Catastrophe for Seabirds.* Washington, DC: ABC. www.abcbirds.org /birdconservationalliance/join/mag_sample2008.pdf.

——. 2007. "International Agreements." http://www.abcbirds.org/conservationiss ues/threats/longlines/international_agreements.html (accessed December 2008).

——. n.d. Albatrosses and Longlining Fact Sheet. http://www.abcbirds.org/conser vationissues/threats/longlines/fact_sheet.html (accessed June 2008).

Anderson, James L. 2003. *The International Seafood Trade*. Boca Raton: CRC Press.

Andresen, Steinar. 2001. "The Convention for the Conservation of Antarctic Marine Living Resources (CCAMLR): Improving Procedures but Lacking Results." In *Environmental Regime Effectiveness*, Miles et al., 405–429. Cambridge: MIT Press.

Andresen, Steinar, Tora Skodvin, Arild Underdal, and Jorgen Wettestad. 2000. *Science and Politics in International Environmental Regimes*. Manchester: Manchester University Press.

Antarctic Treaty System Fact Sheet. 1999. May 27. http://www.asoc.org/general/ats .htm (accessed November 2007).

Ashford, J. R., J. P. Croxall, P. S. Rubilar, and C. A. Moreno. 1994. "Seabird Interactions with Longlining Operations for *Dissostichus eleginoides* at the South Sandwich Islands and South Georgia." *CCAMLR Science* 1: 143–153.

ASOC (Antarctic and Southern Ocean Coalition). 2007. Statement to the Scientific Committee, CCAMLR XXV1, Oct. 24, 2/3. http://www.asoc.org/Home/tabid/36/De fault.aspx (accessed August 2009).

"Attenborough Backs Albatross Campaign." 2005. Press release, October 7. http:// www.4ni.co.uk/news.asp?id=44845 (accessed March 2008).

Australia, Australian Antarctic Division. 2005. *Heard Island and McDonald Islands Marine Reserve Management Plan*. Kingston: AAD. www.heardisland.aq/ (accessed February 2008).

——. n.d.a. "Heard Island and McDonald Islands." http://www.heardisland.aq/ (accessed February 2008).

——. n.d.b. See http://www.aad.gov.au/ for program details (accessed January 2007 onward).

Australia, Australian Nature Conservation Agency. 1996. "Status and Conservation of Albatross and Their Interactions with Fisheries." Final Report. Hobart: Parks and Wildlife Service.

Australia, Commonwealth Scientific and Industrial Research Organization (CSIRO). 2001. "El Niño Link to Southern Ocean Currents." Media release, May 21. www.cs iro.au (accessed June 2006).

——. 2007. "Removing Predators Could Offset Seabird 'Bycatch' Losses." *ScienceDaily*, July 23. http://www.sciencedaily.com/releases/2007/07/070720100041.html (accessed July 2008).

Australia, Customs Service. 2003. "Viarsa Chase." http://www.customs.gov.au/site /page.cfm?u=4691 (accessed June 2007).

Australia, Department of Agriculture, Fisheries and Forestry. 2003. *Seabird Interactions with Longline Fisheries in the Australian Fishing Zone*. Assessment report for the National Plan of Action for Reducing the Incidental Catch of Seabirds in Longline

Fisheries. Canberra: Department of Agriculture, Fisheries and Forestry, http:// www.affa.gov.au/corporate_docs/publications/pdf/fi.

Australia, Department of Environment and Heritage. 2001a. *Macquarie Island Marine Park Management Plan*. Canberra: Environment Australia.

———. 2001b. *Recovery Plan for Albatrosses and Petrels*. Canberra: Environment Australia.

———. n.d. World Heritage, Macquarie Island. deh.gov.au/heritage/ (accessed July 2008).

Australia, Department of Foreign Affairs and Trade. 2005. Treaty between Australia and France, "On Cooperation in the Maritime Areas Adjacent to the French Southern and Antarctic Territories, Heard Island and the McDonald Islands." Canberra, November 24, 2003, entry into force February 1, 2005, *Australian Treaty Series 6*, Australia-France Treaty 2005.pdf.

Australia, Environment Australia. 1998. *Threat Abatement Plan*. Canberra: Biodiversity Group.

———. 2000. Secretariat of the Meeting, *Report on a Meeting to Discuss an Agreement on the Conservation of Southern Hemisphere Albatrosses and Petrels*. Parkes, ACT. www.acap .aq/french/download-document/406–final-report.

Australia, Environment Australia, Natural Heritage Trust. 2001. *Recovery Plan for Albatrosses and Petrels*. Canberra: Natural Heritage Trust.

Australia, Ministry of Fisheries, Forestry and Conservation. 2005. Press release, November 5. mffc.gov.au/releases/2005/05218m.html (accessed June 2006).

Australian Antarctic Magazine 5 (Winter 2003): 15–18.

Baker, G. Barry, R. Gales, S. Hamilton, V. Wilkinson. 2002. "Albatrosses and Petrels in Australia: A Review of Their Conservation and Management." *The Emu* 102: 71–97.

Banks, Joseph. n.d. "Banks's Journal: Daily Entries." The Papers of Sir Joseph Banks at the State Library of New South Wales at http://www2.sl.nsw.gov.au/banks (accessed July 2010).

Baudelaire, Charles. 1857. *The Albatross*. Translated by Richard Howard. poemhunter .com (accessed July 2010).

BBC News. 2005. "MacArthur Gives Support for Albatrosses." April 27. http://news .bbc.co.uk/2/hi/science/nature/448693 (accessed May 2006).

———. 2006. "Celebrities Backing the Albatross Campaign." *Science-Nature*, September 27.

———. 2009. "Closing the Net on Illegal Fishing." February 20. http://search.bbc.co.uk /search?go=toolbar&uri=%2Fsearch%2Ffishing&q=illegal+Fishing (accessed November 2009).

Bell, Brian D., and C. J. R. Robertson. 1994. "Seabirds of the Chatham Islands." In Nettleship, Burger, and Gochfeld 1994, 219–228.

Bialik, Carl. 2009. "How Big Is That Widening Gyre?" *The Wall Street Journal*, March 25.

Biermann, Frank, and Klaus Dingwerth. 2004. "Global Environmental Change and the Nation State." *Global Environmental Politics* 4:1, 1–22.

BirdLife International. 2001. "Crowds Flock to First World Bird Festival." Press release,

June 11. http://www.birdlife.org/news/news/2001/10/510.html (accessed November 2006).

———. 2002a. "America's Cup Teams Pledge Support." Press release, December 23. http://www.birdlife.org/news/news/2002/12/764.html (accessed June 2005).

———. 2002b. "Antarctic Tourists Fund Seabird Conservation." Press release, December. http://www.birdlife.org/news/news/2002/12/764.html (accessed June 2005).

———. 2002c. "Prince of Wales Endorses BirdLife's Save the Albatross Campaign." http://www.birdlife.org/action/campaigns/save_the_albatross/prince_of_wales .html (accessed September 2005).

———. 2002d. "Report of the First South American Workshop on the Conservation of Albatrosses and Petrels." Punta del Este, September, 24–28 (mimeo). http://www .cms.int/pdf/Lima_Punta%20Este%20–%20Albatross%20–%20report%20–%20 final.pdf (accessed November 2007).

———. 2004a. "Celebrities Get Ready for the Ultimate Flutter." Press release, April 27.

———. 2004b. *Tracking Ocean Wanderers: The Global Distribution of Albatrosses and Petrels.* Results from a Tracking Workshop, September 1–5, 2003, Gordon's Bay, South Africa. Cambridge: BirdLife.

———. 2005. "Ecuador Recognises Important Bird Areas." April 15. http://www.bird life.org/news/news/2005/04/ecaudor_ibas.html (accessed May 2007).

———. 2006a. "Prince of Wales Pledges Support." November 5. http://www.birdlife .org/news/news/2006/11/albatross_prince_charles.html (accessed March 2008).

———. 2006b. "Sir David's Personal Plea for Albatrosses." Press release, February 3. http://www.birdlife.org/news/news/2006/02/attenborough.html (accessed March 2008).

———. 2006c. "Skippers Sign Up." Press release, February 20. http://www.birdlife.org /news/news/2006/02/nz_vor.html (accessed June 2007).

———. 2006d. Species fact sheets. http://www.birdlife.org (accessed November 2006 to August 2010).

———. 2007a. "Conservationists Join Ecuador President's Call to Save Galapagos." April 24. http://www.birdlife.org/news/news/2007/04/ecuador_president_gala pagos.html (accessed March 2008).

———. 2007b. Investment in Albatross Conservation Crucial in Tackling New Trawling Threats. http://www.birdlife.org/news/news/2007/06/albatross_investment .htm (accessed August 2010).

———. 2007c. *Sea-Change* (Newsletter), p. 3. http://www.birdlife.org/action/science /species/seabirds/index.html (accessed June 2008).

———. 2008a. Global Seabird Programme. *Albatross Task Force Annual Report 2007.* Sandy: RSPB.

———. 2008b. "Indian Ocean Seabirds Get Thrown a Lifeline." Press release, June 12.

———. 2008c. "Marine Important Bird Areas." http://www.birdlife.org/action/science /sites/marine_ibas/index.html (accessed March 2008).

———. 2008d. "El Programa de Conservación de Aves Marinas en las Américas." http:// www.birdlife.org/action/change/americas/index.html (accessed March 2008).

———. 2008e. "Species Factsheet: Short-tailed Albatross, *Phoebastria albatrus*." http://www.birdlife.org/datazone/speciesfactsheet.php?id=3956 (accessed March 2009).

———. 2008f. "Trawling a Major Threat to Seabirds." http://www.rspb.org.uk/news/details.asp?id=tcm:9–196058 (accessed April 2009).

———. 2008g. "Which Species Are Affected?" http://www.birdlife.org/action/campaigns/save_the_albatross/species_list.html (accessed December 2008).

———. 2008h. "Working Together for Birds and People." http://www.birdlife.org/index.html (accessed March 2008).

———. 2010a. "Albatross Species Fact Sheet." www.birdlife.org/action/campaigns/save_the_albatross/species_list.html (accessed July 2010).

———. 2010b. "Albatross Task Force" diaries. http://www.birdlife.org/community2010/10/albatross-task-force-diaries—-juliano-cesar-in-brazil/ (accessed August 2010).

———. 2010c. "Species at Risk," under Save the Albatross Campaign. http://www.rspb.org.uk/supporting/campaigns/albatross/about/species/index.asp (accessed August 2010).

Birkhead, Tim. 2008. *The Wisdom of Birds*. London: Bloomsbury.

Bjordal, A., and S. Lokkeborg. 1996. *Longlining*. Oxford: Fishing News.

Blakely, Laurence. 2008. "End of the *Viarsa* Saga." *Pacific Rim Law & Policy Journal* 117 (June): 677. www.law.washington.edu/pacrim/.../Default.aspx?a=20080601703 (accessed June 2009).

Blue Ocean Institute. n.d. "Seafood, albatross, longlining, etc." http://www.blueocean.org/home (accessed June 2006 and onward).

Bonner, W. N., and D. W. H. Walton. 1985. *Antarctica*. Oxford: Pergamon.

Bradsher, Keith. 2005. "Trade Ministers at W.T.O. Talks Agree to Limit Fishing Subsidies." *New York Times*, December 15.

British Ornithologists' Union. 1971. *The Status of Birds in Britain and Ireland*. Oxford: Blackwell.

Brody, Jane E. 2004. "In the Land of Kiwis." *New York Times*, March 23.

Brooke, Michael. 2004. *Albatrosses and Petrels Across the World*. Oxford: Oxford University Press.

Brothers, Nigel. 1991. "Albatross Mortality and Associated Bait Loss in the Japanese Longline Fishery in the Southern Ocean." *Biological Conservation* 55: 255–268.

———. 1996a. *Catching Fish, Not Birds: A Guide to Improving Your Longline Fishing Efficiency*. Hobart: Australian Parks and Wildlife Service.

———. 1996b. *Longline Fishing: Dollars and Sense*. Hobart: Australian Parks and Wildlife Service.

Brothers, Nigel P., John Cooper, and Svein Lokkeborg. 1999. *The Incidental Catch of Seabirds by Longline Fisheries: Worldwide Review and Technical Guidelines for Mitigation*. Rome: FAO.

Brothers, Nigel, and Andrew Foster. 1997. "Seabird Catch Rates." *Marine Ornithology* 25: 37–42.

Brothers, Nigel, and Rosemary Gales. 1996. "Status and Conservation of Albatross."

Australian Conservation Agency. Hobart: Australian Conservation Agency, mimeo.

Brown, Derek, et al. 2001. "Report on the Falklands Conservation Rat Eradication Project." *Falklands Conservation*. Stanley: Falklands Conservation. http://www.falk landsconservation.com/wildlife/conservation_issues/rat_eradication-final.html (accessed March 2008).

Brown, J. Stanley. 1894. "Fur Seals and the Bering Sea Arbitration." *Journal of the American Geographical Society of New York* 26: 44–47.

Brown, Paul. 2000. "UN to Crack Down as Pirate Boats Threaten to Drive Fish to Extinction." *The Guardian*, August 15.

Buck, Susan J. *The Global Commons*. Washington, DC: Island Press, 1998.

Burgmans, Antony. 2008. "Antony Burgmans." http://www.elida.co.uk/ourcompany /aboutunilever/companystructure/nonexecutivedirectors/antony_burgmans. asp and www.referenceforbusiness.com/.../A.../Burgmans-Antony-1947.html (accessed March 2008).

Burhenne-Guilmin, F. 2004. "How CMS was Born." In CMS 2004, 19.

Carpenter, Eric. 2004. "Seabirds of the Lower Texas Coast." http://emyadestes.com /lrgv_p.

elagic.htm; http://listserv.uh.edu/cgibin/wa?A2=indo310&L=texbirds&T=o&O=D& P=10186,http://www.texasbirds.org/tbrc/ar2003.html (accessed December 2007).

Casey, Michael. 2005. "Bird Flies Nearly 2,500 Miles for Baby's Food." *Associated Press*, December 24, in news-press.com.

CBD (Convention on Biological Diversity). 2010. "The Rio Conventions." http://www .cbd.int/2010 (accessed December 2010).

CCAMLR (Convention for the Conservation of Antarctic Marine Living Resources). 1994. *Report of the Thirteenth Meeting of the Commission*, Item 4. http://www.ccamlr .org/pu/E/e_pubs/cr/94/toc.htm (accessed April 2007).

———. 2005. "CCAMLR Symposium." Valdivia, Chile, Vol. 2. www.aad.gov.au/.../ml _386285141319444_CCAMLR%20Symposium%20Volume%202.pdf (accessed June 2009).

———. 2007. "Work on the Elimination of Seabird Mortality Associated with Fishing." *www.ccamlr.org/pu/E/sc/imaf/.../CCAMLR_elimination%20of%20IMAF.pdf*.

———. 2009a. Report of the Twenty-eighth Meeting of the Commission, Items 8 and 9. http://www.ccamlr.org/pu/e/e_pubs/cr/drt.htm (accessed July 2010).

———. 2009b. "Schedule of Conservation Measures." History Table. http://www .ccamlr.org/pu/e/e_pubs/cm/08–09/toc.htm (accessed July 2010).

———. n.d. General information, documents, commission meetings, scientific committee reports, conservation measures, etc. http://www.ccamlr.org/pu/e/gen-intro.htm (accessed July 2005 and onward).

CCSBT (Commission for the Conservation of Southern Bluefin Tuna). 2005. "Report of the Extended Scientific Committee for the Tenth Meeting of the Scientific Committee." September 5–8, Taipei, Taiwan, Appendix 2, Agenda Item 4, "Review of

SBT Fisheries" and Working Group on Ecologically Related Species. http://www
.ccsbt.org/index.html (accessed December 2006 and March 2008).

Chasek, Pamela S. 2001. *Earth Negotiations*. Tokyo: United Nations University Press.

Childerhouse, Simon, Christopher Robertson, Wally Hockly, and Nadine Gibbs.
2003. "Royal Albatross on Enderby Island." New Zealand, DOC, Supplement to
Department of Conservation, Science Internal Series, 144. www.doc.govt.nz/up
load/documents/science-and.../dsis144.pdf.

CIMAS (Cooperative Institute for Marine and Atmospheric Studies). n.d. "Ocean Surface Currents." http://oceancurrents.rsmas.miami.edu/ (accessed July 2006).

CITES (Convention on International Trade in Endangered Species of Wild Fauna
and Flora). 2002a. "Conservation of and Trade in *Dissostichus* Species," CoP (Conference of the Parties) 12 Doc. 44, Santiago, Chile, November 3–15. http://www
.googlesyndicatedsearch.com/u/cites?q=patagonian+toothfish&submit=Go (accessed March 2008).

———. 2002b. "Cooperation between CITES and the Commission for the Conservation
of Antarctic Marine Living Resources regarding Trade in Toothfish." CoP12 Doc.
4, and (later) CoP13 Doc. 36. http://www.googlesyndicatedsearch.com/u/cites
?q=toothfish&submit=Go (accessed September 2007 and onward).

———. n.d. Resources about Tuna, Toothfish. http://www.cites.org/ (accessed February 2002 and onward).

Clover, Charles. 2004. *The End of the Line*. London: Eubury Press.

CMS (Convention on the Conservation of Migratory Species of Wild Animals; also
called Bonn Convention). 2004. *25 Years of Journeys*. Nairobi: UNEP (United Nations
Environment Programme). www.cms.int/publications/index.htm (accessed June
2007).

———. 2010. "Parties to the Convention." www.cms.int/about/part_lst.htm (accessed
December 2010).

———. n.d. "Appendix 1 & 11 of CMS." http://www.cms.int/documents/appendix/cms_
app1_2.htm (accessed September 2005).

Collier, Graham. 1999. *Antarctic Odyssey*. New York: Carroll and Graf.

Collier, Paul. 2007. *The Bottom Billion*. New York: Oxford University Press.

COLTO (Coalition of Legal Toothfish Operators). 2006a. "Governments Make Recommendations to Help Stop Pirate Fishing." Media release, March 3. http://www
.colto.org/ (accessed March 2007).

———. 2006b."Greenpeace UK, 'Make Piracy History.'" Media release, February 28.
http://www.colto.org/ (accessed March 2008).

———. 2007. "South Georgia: Fishing and Shipping News." October 10. http://www
.colto.org/Articles_Latest.htm (accessed March 2008).

———. 2009. "Toothfish Vessels." http://www.colto.org/Toothfish_Vessels.htm (accessed March 2009).

Conacher, Arthur, and Jeanette Conacher. 2000. *Environmental Planning and Management in Australia*. Melbourne: Oxford University Press.

Constable, Andrew. 2001. "The Status of Antarctic Fisheries Research." eprints.utas .edu.au/2661/15/14_Constable.pdf.

Corson, Trevor. 2007. *The Zen of Fish: The Story of Sushi, from Samurai to Supermarket.* New York: HarperCollins.

COSEWIC (Committee on the Status of Endangered Wildlife in Canada). 2003. *Assessment and Status Report on the Short-tailed Albatross* Phoebastria albatrus *in Canada.* Ottawa: www.sararegistry.gc.ca/status/status_e.cfm (accessed June 2004).

Croxall, J. P. 1979. "Distribution and Population Changes in the Wandering Albatross *Diomedea exulans* L. at South Georgia." *Ardea* 67: 15–21.

———. 1984. "Status and Conservation of the World's Seabirds." In Croxall, Evans, and Schreiber 1984, 2–11.

———. 1990a. "Guest Editorial." *Antarctic Science* 2:1 (March): 1.

———. 1990b. "Impact of Incidental Mortality on Antarctic Marine Vertebrates." *Antarctic Science* 2:1 (March): 1.

———. 2003. "Minutes of Evidence." House of Lords, Science and Technology Committee, Sub-Committee 1, International Agreements, December 2. London: Houses of Parliament, UK, 39 pp.

———. 2004a. "Godman-Salvin Medal." *Ibis* 146: 549–550.

———. 2004b. "Professor John Patrick Croxall." *London Gazette,* June 12, p. B7. http://www.london-gazette.co.uk (accessed February 2007).

———. 2008a. Interview, May 29, at BirdLife International in Girton, Cambridge, UK.

———. 2008b. "The Role of Science and Advocacy in the Conservation of Southern Ocean Albatrosses." *Bird Conservation International* 18: 1–17.

Croxall, J. P., P. G. H. Evans, and R. W. Schreiber, eds. 1984. *Status and Conservation of the World's Seabirds.* ICBP Technical Publication No. 2. Cambridge: ICBP.

Croxall, J. P., and Rosemary Gales. 1998. "An Assessment of the Conservation Status of Albatrosses." In Robertson and Gales 1998a, 46–65.

Currie, Duncan. 2005. "Deep, Deep Trouble." Greenpeace International, March 15. http://www.greenpeace.org/international/press/reports/deep-deep-trouble (accessed March 2007).

Dalziell, J., and M. de Poorter. 1993. "Seabird Mortality in Longline Fisheries around South Georgia." *Polar Record* 29: 143–145.

Davis, Bruce W. 1996. "Contemporary Ocean and Coastal Management Issues in Australia and New Zealand: An Overview." *Ocean & Coastal Management* 33: 5–18.

Deguchi, Tomohiro, and Yamashina Institute for Ornithology. 2010. "Translocated Short-tailed and Black-tailed Albatrosses Continue to Do Well." June 23. http://www.acap.aq/latest-news/ (accessed June 2010).

Dingwall, P. R., ed. 1995. *Progress in the Conservation of Sub-Antarctic Islands.* Cambridge: IUCN Publications.

Double, Mike. 2007. "Potential of Blue-dyed Bait." In Third Meeting of Advisory Committee, Seabird Bycatch Working Group, June 19–22, ACAP, AC3 Doc. 14 Rev. 5, Agenda Item 12, p. 8. http://www.acap.aq/en/index.php?option=com_docman&task=cat_view&gid=66&Itemid=33 (accessed February 2008).

Doughty, Robin W. 1975. *Feather Fashions and Bird Preservation: A Study in Nature Protection*. Berkeley: University of California Press.

———. 2010. "Saving the Albatross: Fashioning an Environmental Regime." *Geographical Review*, Special Issue on Avian Geography, 100:2 (April): 216–228.

Dovers, Stephen, ed. 1994. *Australian Environmental History*. Melbourne: Oxford University Press.

Duffy, David C. 1994. "Toward a World Strategy for Seabird Sanctuaries." *Colonial Waterbirds* 17: 200–206.

Dunmore, John. 2006. *Mrs. Cook's Book of Recipes for Mariners in Distant Seas*. http://www.panmacmillan.com.au/display_title.asp?ISBN=9780908988648&Author=Dunmore,%20John (accessed June 2010).

Dutter, Barbie. 2006. "Antarctic 'Cold Rush' Raises Fears." *The Sunday Telegraph* (UK), June 4.

Edwards, George. 1747. *A Natural History of [Uncommon] Birds, Most of which have not been figured or described, . . .* (usually referred to as *Uncommon Birds*). 7 vols. London: Royal College of Physicians.

Elliott, MacAlister, and Partners. 2003. "Impact of Fisheries Bycatch on Endangered Migratory Species." Final Report, CRO278, UK, Department of Environment and Rural Affairs. Lymington: Elliott. www.defra.gov.uk/marine/index.htm (accessed July 2006).

Elliott, Valerie. 2004. "Wandering Albatrosses Soar Out of the Race." *The Times* (London), July 1.

Environmental Treaties and Resource Indicators (ENTRI). n.d. http://sedac.ciesin.columbia.edu/entri/texts/protection.of.birds.1950.html (accessed February 2008).

Falklands Conservation. 2004. *Falklands National Plan of Action for the Reduction of the Incidental Catch of Seabirds in Longline Fisheries*. Falkland Islands.

———. 2009. *Falklands Plan of Action for Reducing Incidental Catch of Seabirds in Trawl Fisheries*. Falkland Islands. http://www.falklandsconservation.com/.

———. n.d. "Report." Stanley: Falklands Conservation. http://www.falklandsconservation.com/SeabirdRepo304.html (accessed March 2008).

Fallon, Liza D., and Lorne K. Kriwoken. 2004. "International Influence of an Australian Nongovernment Organization in the Protection of Patagonian Toothfish." *Ocean Development & International Law* 35:3, 221–266.

Farr-Biswell, Shelly. 2009. "Mitigation in Motion: Saving Seabirds at Sea." *Seafood New Zealand*. 17:3 (April): 16–19.

Fickling, David. 2003. "Australian and South African Officials Board Suspect Boat." *The Guardian*, August 28.

Fikkan, Anne, et al. 1993. "Polar Bears." In Young and Osherenko 1993, 96–151.

Fisher, Harvey. 1966a."Airplane-Albatross Collisions on Midway Atoll." *The Condor* 68:3 (May–June): 229–242.

———. 1966b. "Midway's Deadly Antennas." *Audubon Magazine* 68: 220–223.

———. 1970. "The Death of Midway's Antennas." *Audubon Magazine* 72: 62–63.

Fisher, James. 1952. *The Fulmar*. London: Collins.

Fisher, Mildred L. 1970. *The Albatross of Midway Island: A Natural History of the Laysan Albatross*. Carbondale: Southern Illinois University Press.

Fishers Forum. 2006. *Report of South American Fishers Forum*. Sao Paulo, December. Other entries are also at http://www.fishersforum.net (accessed October 2008); www.southernseabirds.org/f1024,45694/45694_SAFF_Report.pdf (accessed December 2008).

Fitzmaurice, Malgosia. 2004. "The International Court of Justice and the Environment." *Non-State Actors and International Law* 4: 191.

Fleming, C. A. 1984. "Lancelot Eric Richdale." *Proceedings of the Royal Society of New Zealand* 112 (June): 27–37.

Forest & Bird. 2001a. "Forest & Bird to Launch Save the Albatross Campaign in New Zealand." December 13. http://www.forestandbird.org.nz/mediarelease /1997_2003/03index.asp (accessed February 2007).

———. 2001b. "Get Involved with the Oceans Debate." *Conservation News*, August.

———. 2001c. "A Sea Change." Press release, December 28.

———. 2005. Press release, March 8. http://www.forestandbird.org.nz/search/node /Media+releases (accessed July 2007).

———. 2006. "Review of the New Zealand Coastal Policy Statement." http://www .forestandbird.org.nz/search/node/Media+releases (accessed July 2007).

———. 2007a."Fishing Industry Gets Subsidy." Press release, April 4. (http://www .forestandbird.org.nz/what-we-do/publications/media-releases/fishing-industry- gets-24–million-subsidy-'giving-up'-nothing (accessed July 2009).

———. 2007b. "Hoki Fishery Doesn't Deserve Its 'Sustainability Tick.'" Media release, September 17. http://www.forestandbird.org.nz/mediarelease/2007/0917 (accessed June 2008).

———. 2007c. "Save the Albatross." http://www.forestandbird.org.nz/Marine/albatross .asp (accessed March 2008).

———. 2010. "Leading Conservation Groups Unite on World Oceans Day." Press release, June 7. http://www.forestandbird.org.nz/what-we-do/publications/media- releases/leading-conservation-groups-unite-on-world-oceans-day (accessed August 2010).

Fourth International Fishers Forum. 2007. "IFF4 Mission and Objectives." www .fishersforum.net/IFF4/IFF4%20agenda%20D3_draft%20final_web.pdf (accessed March 2007).

Fox, Rebecca. 2008. "Ranger Helps Japanese Albatross." *Otago Times*, June 21.

Franzen, Jonathan. 2010. "Emptying the Skies." *The New Yorker*, July 26.

Frawley, Kevin. 1994. "Evolving Visions: Environmental Management and Nature Conservation in Australia." In Dovers 1994, 55–78.

Gales, Rosemary. 1993. *Co-Operative Mechanisms for the Conservation of Albatross*. Hobart: Australian Antarctic Foundation.

———. 1998. "Albatross Populations: Status and Threats." In Robertson and Gales 1998a, 20–45.

Gardner, Martin. 1965. *The Annotated Ancient Mariner: The Rime of the Ancient Mariner*. London: Blond.

Gaston, A. J. 2004. *Seabirds: A Natural History*. New Haven: Yale University Press.

Gianni, Matthew. 2002. "The Entry into Force of the 1995 UN Fish Stocks Agreement: An NGO Perspective." *Reporter* 2 (January/February); reprinted in Ocean Law-Online, Paper 8. www.oceanlaw.net/projects/projects3.html (accessed October 2006).

Gianni, Matthew, and Walt Simpson. 2005. *The Changing Nature of High Seas Fishing*. Australian Department of Agriculture, Fisheries and Forestry, International Transport Workers' Federation, and WWF International. N.p.

"Giant Tuna Fetches 111,000 [pounds sterling] at Tokyo Auction." 2010. *Guardian.co.uk* January 5. http://www.guardian.co.uk/environment/2010/jan/05/giant-tuna-toky -auction (accessed July 2010).

Gibson, J. D. 1963. "Third Report of the New South Wales Albatross Study Group (1962) Summarizing Activities to Date." *The Emu* 63:3 (November): 215–223.

———. 1967. "The Wandering Albatross (*Diomedea exulans*): Results of Banding and Observations in New South Wales Coastal Waters and the Tasman Sea." *Notornis* 14 (June): 47–57.

Gibson, J. D., and A. R. Sefton. 1955. "Notes on Some Albatrosses Off Coastal New South Wales." *The Emu* 55:1 (March): 44–48.

———. 1959. "First Report of the New South Wales Albatross Study Group." *The Emu* 59 (May): 73–82.

———. 1960. "Second Report of the New South Wales Albatross Study Group." *The Emu* 60 (May): 125–130.

Gilman, Eric, Christopher H. Boggs, and Nigel Brothers. 2003. "Performance Assessment of an Underwater Setting Chute . . . in the Hawaiian Pelagic Longline Tuna Fishery." *Ocean & Coastal Management* 46:11–12, 985–1010.

Gilman, Eric, and Holly Freifeld. 2003. "Introduction to Seabird Mortality in North Pacific Longline Fisheries." *Endangered Species Update* 20:2 (March): 35.

Greenberg, Paul. 2010a. *Four Fish: The Future of the Last Wild Food*. New York: Penguin.

———. 2010b. "Tuna's End." *New York Times Magazine*, June 27, 28–37, 44–48.

Greenpeace International. 2001. *Pirate Fishing: Plundering the Oceans*. N.p.: Greenpeace.

———. 2007. Resource for seabird and pirate rishing. http://www.greenpeace.org/in ternational/campaigns/oceans (accessed May 2007 and onward).

———. n.d. "Patagonian Toothfish." http://archive.greenpeace.org/oceans/piratefish ing/toothfish.html (accessed August 2008).

Hall, Martin A., D. L. Alverson, and K. I. Metuzals. 2000. "By-Catch: Problems and Solutions." *Marine Pollution Bulletin* 41: 204–219.

Hannon, Robert. 2006. Arctic Science Journeys Radio, Fairbanks, Alaska.

Hansom, James D., and John E. Gordon. 1998. *Antarctic Environments and Resources*. New York: Longman.

Hardin, Garrett. 1968. "The Tragedy of the Commons." *Science* 162, 1243–1248.

Harris, U., and A. Orsi. 2001 (updated 2006). "Locations of the Various Fronts in the Southern Ocean." Australian Antarctic Data Centre—CAASM Metadata (http://data.aad.gov.au/aadc/metadata/). Direct link to metadata is http://data.aad.gov.au/aadc/metadata/metadata_redirect.cfm?md=AMD/AU/southern_ocean_fronts (accessed August 2010).

Harrison, Neil E., and Gary C. Bryner, eds. 2004. *Science and Politics in the International Environment*. Lanham: Rowman and Littlefield.

Hasegawa, Hiroshi. 1984. "Status and Conservation of Seabirds in Japan, with Special Attention to the Short-tailed Albatross." In Croxall, Evans, and Schreiber 1984, 487–500.

Haward, Marcus, and A. Bergin. 2001. "The Political Economy of Japanese Distant Water Tuna Fisheries." *Marine Policy* 25, 91–101.

Haward, Marcus, A. Bergin, and H. Robert Hall. 1998. "International Legal and Political Bases to the Management of the Incidental Catch of Seabirds." In Robertson and Gales 1998a, 255–266.

Headland, Robert. 1984. *The Island of South Georgia*. New York: Cambridge University Press.

Henderson, Mark. 2004. "Have a Flutter." *The Times* (London), January 23.

Herr, Richard. 2001. "The International Regulations of Patagonian Toothfish: CCAMLR and High Seas Fisheries Management." In Stokke 2001b, 303–328.

Herr, Richard A., and Bruce W. Davis. 1996. "ATS Decision-making and Change: The Role of Domestic Politics in Australia." In Stokke and Vidas 1996, 331–360.

Higham, J. E. S. 1998. "Tourists and Albatrosses." *Tourism Management* 9: 521–531.

High Seas Task Force. 2006. *Closing the Net: Stopping Illegal Fishing on the High Seas*. March 3. http://www.illegal-fishing.info/item_single.php?item=document&item_id=58&approach_id=8.

Hill, Senator Robert. 1997. Government Office, press release, April 16.

Hindwood, K. A. 1955. "Sea-Birds and Sewage." *The Emu* 55 (September): 212–216.

Hoshaw, Lindsey. 2009. "Afloat in the Ocean." *New York Times*, November 10.

Howkins, Adrian J. 2008. "Frozen Empires: A History of the Antarctic Sovereignty Dispute between Britain, Argentina, and Chile, 1939–1959." Ph.D., diss., University of Texas at Austin.

Hutchison, Kristan. 2004. "Fighting Over Fish." *The Antarctic Sun*, February 2, 1 and 16. http://antarcticsun.usap.gov/pastIssues/ (accessed August 2007).

ICCAT (The International Commission for the Conservation of Atlantic Tunas). 2007a. "About ICCAT." http://www.iccat.int/introduction.htm (accessed March 2008).

———. 2007b. "Report of the Sub-Committee on Ecosystems." February 19–23, Madrid, 3–7, SC-ECO-REPORTALL-15032007.pdf. http://www.iccat.int/search.asp?zoom_query=seabird+bycatch&zoom_per_page=10&zoom_and=1&zoom_sort=0 (accessed March 2008).

"Illegal Fishing in the Southern Ocean." 2003. *Australian Antarctic Magazine* 5 (Winter). http://www.aad.gov.au/default.asp?casid=11981.

"Impacts of Fishing Techniques." http://www.lighthouse-foundation.de/index .php?id=94&L=1 (accessed February 2008).

International Association of Antarctic Tour Operators. 2006. *Seabird Conservation Newsletter*, 1–6. http://www.iaato.org/info.html (accessed June 2008).

——. 2007. Directory. http://apps.iaato.org/iaato/directory/ (accessed December 2007).

International Seafood Sustainability Foundation. 2009. "Bycatch in the World's Tuna Fisheries." www.iss-foundation.org/FileContents.phx?fileid=566fad9f-4e52–483a (accessed August 2010).

International Tuna Fishers Conference on Responsible Fisheries and Third International Fishers Forum. 2005. fishersforum.net/old/IFF3/YokohamaDeclaration .pdf.

Invasive Species in the UK Overseas Territories. 2007. www.scscb.org/working_ groups/resources/invasive-species-in-the-ukots.doc (accessed March 2008).

IOTC (Indian Ocean Tuna Commission). 2007. "Report of the Third Session Working Party on Ecosystems and Bycatch." IOTC-2007–WPEB- R[E], p. 3, July 11–13, Seychelles. http://www.iotc.org/English/documents/doc_proceed_details.php?docid =1241 (accessed March 2008 and onward).

Issenberg, Sasha. 2007. *The Sushi Economy: Globalization and the Making of a Modern Delicacy*. New York: Penguin.

IUCN (International Union for the Conservation of Nature). 2010. Red List of Threatened Species. http://www.iucnredlist.org/ (accessed July 2010).

Jameson, William. 1961. *The Wandering Albatross*. Garden City: Doubleday.

Japan, Fisheries Agency. 2001 [2005]. *Japan's National Plan of Action for Reducing Incidental Catch of Seabirds in Longline Fisheries*. Tokyo: Fisheries Agency. ftp://ftp.fao.org/FI /DOCUMENT/IPOAS/national/japan/NPOA-seabirds.pdf (accessed March 2008).

Johnson, Paul A., and David Santillo. 2004. "Conservation of Seamount Ecosystems." *Archive of Fishery and Marine Research* 51: 305–319.

Jolly, David. 2010. "Certification of Krill Harvest." *New York Times*, June 23.

Jouventin, P., and J. P. Roux. 1983. "Discovery of a New Albatross." *Nature* 305: 181.

Jouventin, Pierre, J. C. Stahl, H. Weimerskirch, and J. L. Mougin. 1984. "The Seabirds of the French Subantarctic Islands & Adelie Land, Their Status and Conservation." In Croxall, Evans, and Schreiber 1984, 609–625.

Jouventin, Pierre, and Henri Weimerskirch. 1990. "Satellite Tracking of Wandering Albatrosses." *Nature* 343 (February 22): 746–749.

——. 1991. "Changes in the Population Size and Demography of Southern Seabirds: Management Implications." In Perrins, Lebreton, and Hirons 1991, 297–314.

Joyner, Christopher C. 1998. *Governing the Frozen Commons*. Columbia: University of South Carolina Press.

——. 2005. *International Law in the 21st Century*. Lanham: Rowan and Littlefield.

Juda, Lawrence. 1996. *International Law and Ocean Use Management*. London: Routledge.

Kock, Karl-Hermann. 1992. *Antarctic Fish and Fishes*. New York: Cambridge University Press.

———. 1994. "Fishing and Conservation in Southern Waters." *Polar Record* 30: 3–22.

———. 2001. "The Direct Influence of Fishing and Fishery-Related Activities on Non-Target Species in the Southern Ocean." *Reviews of Fish Biology and Fisheries* 11: 31–56.

Kriwoken, L. K., and D. M. Rootes. 2000. "Tourism on Ice: Environmental Impact Assessment and Antarctic Tourism." *Impact Assessment and Project Appraisal* 18: 138–150.

Kwok, Kai Yin, Cynthia Yau, and I-Hsun Ni. 2002. "Conservation Aspects of Commercial Fishing." *Proceedings* of IUCN/WCPA-EA-4 Taipei Conference, March 18–23, Taipei, Taiwan, 377.

Lack, David. 1965. *Enjoying Ornithology*. London: Methuen.

Lack, Mary. 2008. *Continuing CCAMLR's Fight against IUU Fishing for Toothfish*. N.p.: WWF Australia and TRAFFIC International.

Lack, M., and G. Sant. 2001. "Patagonian Toothfish: Are Conservation and Trade Measures Working?" *Traffic Network*. http://www.traffic.org/toothfish/tooth5.html (accessed May 2006).

Ladbrokes. 2005."Big Bird Race." http://www.ladbrokes.com/bigbirdrace/ (accessed January 2007).

Lawrence, Felicity. 2006. "The Fish List: How Your Supermarket Rates." *The Guardian*, March 1.

Lewis Smith, R. I. 1995. "Conservation Status of South Georgia and the Sandwich Islands." In Dingwall 1995, 3–14.

Liermann, Charles. 1999. *Depensation: Evidence and Implications*. Seattle: University of Washington Press.

Lighthouse Foundation. n.d. "Impacts of Fishing Techniques on the Patagonian Toothfish Fishery in the Southern Ocean." http://www.lighthouse-foundation.de/index.php?id=94&L=1 (accessed February 2008).

Liittschwager, David, and Susan Middleton. 2005. "Hawaii's Outer Kingdom." *National Geographic* (October). http://ngm.nationalgeographic.com/2005/10/hawaii/liittschwager-middleton-text (accessed December 4, 2010).

Lindsey, Terence. 2008. *Albatrosses*. Melbourne: CSIRO.

Lloyd, Clare, M. L. Tasker, and K. Partridge. 1991. *The Status of Seabirds in Britain and Ireland*. London: Poyser.

Lockley, Ronald. 1942. *Shearwaters*. London: Dent.

———. 1953. *Puffins*. London: Dent.

Lockley, Ronald, and James Fisher. 1954. *Seabirds*. London: Collins.

Lokkeborg, S., and W. Thiele. 2004. Report of the FAO/BirdLife South American Workshop on Implementation of NPOA-Seabirds and Conservation of Albatrosses and Petrels. Valdivia, Chile, December 2–6, 2003. Rome: FAO. http://www.fao.org/docrep/007/y5742e/y5742e00.htm (accessed March 2008).

Lyster, Simon. 1985. *International Wildlife Law*. Cambridge: Grotius Publications.

Macquarie Island World Heritage Area. n.d. www.parks.tas.gov.au/Macquarie/mac qis.html (accessed February 2008).

Majkowski, Jacek. 1998. *Global Fishery Resources of Tuna and Tuna-like Species*. FAO Fisheries Technical Paper 483. Rome: FAO.

Marchant, S., and P. J. Higgins. 1990. *Handbook of Australian, New Zealand and Antarctic Birds*. Vol. 1: *Ratites to Ducks*. Melbourne: Oxford University Press.

Marine Resources Assessment Group. 2005a. www.onefish.org/.../marine_resources _assessment_group_mrag_uk_en_225713_114910.htm.

———. 2005b. *IUU Fishing on the High Seas: Final Report*. London: Dept. for International Development. www. high-seas.org HSTF, Documents (accessed April 2007).

Marshall, Leon. 2004. "Model Jerry Hall's Albatross Wins the Big Bird Race." *National Geographic News*, July 13.

Masefield, John. "Sea Change." www.poet.com/masefield-john/sea-change.html (accessed February 2008).

Matthews, Leonard Harrison. 1929. "The Birds of South Georgia." *Discovery Reports* 1:563–592.

———. 1931. *South Georgia: The British Empire's Sub-Antarctic Outpost*. Bristol: Wright.

———. 1951. *Wandering Albatross: Adventures among the Albatrosses and Petrels in the Southern Ocean*. London: Macgibbon and Kee.

McConigal, D., and L. Woodworth. 2001. *Antarctica and the Arctic: The Complete Encyclopedia*. Willowdale, Ontario: Firefly.

McCurry, Justin. 2007. "Tuna Stocks Close to Exhaustion Says WWF." *The Guardian*, January 2. http://www.guardian.co.uk/environment/2007/jan/23/fishing.conser vation (accessed April 2009).

McDaniel, James W. 1973. "Vagrant Albatrosses." *American Birds* 27 (June): 563–565.

Medway, David. 1998. "Human-Induced Mortality of Southern Albatrosses." In Robertson and Gales 1998a, 189–198.

Meikle, James. 2006. "Illegal Fishing Worth $9bn a Year." *The Guardian*, March 4.

Melvin, Edward F. 2000. "Streamer Lines to Reduce Seabird Bycatch in Longline Fisheries." Seattle: Washington Sea Grant Program. http://www.wsg.washington.edu /communications/online/streamers (accessed August 2010).

Melvin, Edward F., and Julia K. Parrish, eds. 2001. *Seabird Bycatch Trends, Roadblocks, and Solutions*. Fairbanks: University of Alaska Sea Grant College Program.

Mier, K. L., et al. 2000. *Conservation of Marine Habitats in the Region of Heard and McDonald Islands*. Report to Environment Australia. Kingston: AAD.

Miles, Edward L., et al. 2002. *Environmental Regime Effectiveness*. Cambridge: MIT Press.

Mitchell, Ronald B. 2003. "International Environmental Agreements." *Annual Review of Environmental Resources* 28: 216429–216461.

Mlodinow, Steven G. 1999. "Southern Hemisphere Albatrosses in North American Waters." *Birders Journal* 8:3 (June/July): 131–141.

Molenaar, Erik J. 2001. "CCAMLR and Southern Ocean Fisheries." *The International Journal of Marine and Coastal Law* 16:3, 465–499.

Montevecchi, William A., 2002. "Interactions between Fisheries and Seabirds." In

Biology of Marine Birds, ed. E. A. Schreiber and Joanna Burger, 527–557. Boca Raton: CRC.

Morling, Paul. 2004. "The Economics of Marine Protected Areas in the High Seas." *George Wright Forum* 21: 4, 49–61.

MSC (Marine Stewardship Council). 2002. "Vision Mission Values." http://www.msc .org/html/content_482.htm (accessed March 2007).

———. 2003."Whole Foods Market's Supports MSC Fishery Outreach." http:// www.msc.org/newsroom/news/whole-foods-markets-supports-msc-fishery-outreach?searchterm=whole+foods (accessed August 2009).

———. 2004. "Objections Panel Concludes Review." http://www.msc.org/html/ni_120 .htm (accessed March 2008).

———. 2006. "Wal-Mart Stores, Inc. Introduces New Label to Distinguish Sustainable Seafood."http://www.msc.org/newsroom/news/wal-mart-stores-inc.-introduces -new-label-to/?searchterm=Walmart (accessed August 2009).

———. 2007. "MSC Response to Marks & Spencer Announcement on Sustainable Fish." January 15. http://www.msc.org/newsroom/news/msc-response-to-marks-spencer-announcement-on/?searchterm=marks%20and%20spencer (accessed August 2010).

———. 2009. "South Georgia Patagonian Toothfish Longline." [Certified in 2004 and in 2009.] http://www.msc.org/track-a-fishery/certified/south-atlantic-indian-ocean/south-georgia-patagonian-toothfish-longline/?searchterm=patagonianto othfish (accessed December 2010).

———. 2010a. "Aker BioMarine Krill Objection Concluded." May 26. http://www.msc .org/newsroom/news/aker-biomarine-krill-objection-concluded-certifiers-de termination-confirmed (accessed June 2010).

———. 2010b. "Certified Fisheries." http://www.msc.org/track-a-fishery/certified (ac-cessed December 2010).

———. 2010c. "Ross Sea Toothfish Objection." May 10. http://www.msc.org/news room/news/ross-sea-toothfish-objection-certifier-asked-to-reconsider-its-scor ing-and-give-further-review-to-stakeholder-comments (accessed July 2010).

———. n.d.a. "Americas: Market Overview." http://www.msc.org/newsroom/news /msc-ecolabelled-products-crosses-2–000–milestone?searchterm=Uk+superma rkets+2006 (accessed August 2009).

———. n.d.b. "History and Timeline." http://www.msc.org/html/content_470.htm (ac-cessed June 2007 and onward).

———. n.d.c. "How the MSC Is Funded." http://www.msc.org/html/content_2036.htm (accessed July 2009).

Muffet, William Carroll. 2000. "The International Protection of Wildlife." In *Interna-tional, Regional, and National Environmental Law*, ed. Fred L. Morrison and R. Wol-frum, 343–354. Hague: Kluwer.

Murphy, Robert C. 1936. *Oceanic Birds of South America*. 2 vols. New York: American Museum of Natural History.

———. 1965. *Logbook for Grace: Whaling Brig Daisy, 1912–1913*. New York: Time.

Myers, Ransom A., and Boris Worm. 2003. "Rapid Worldwide Depletion of Predatory Fish Communities." *Nature* 423 (May 15): 280–283.

National Environmental Trust. 2004. *Black Market for White Gold: The Illegal Trade in Chilean Sea Bass.* Washington, DC: National Environmental Trust.

National Geographic Society. 2007. "Still Waters." *National Geographic* 211: 4, 61.

Naughton, M. B, M. D. Romano, and T. S. Zimmerman. 2007. "A Conservation Action Plan for Black-footed Albatross (*Phoebastria nigripes*) and Laysan Albatross (*P. immutabilis*)," Ver. 1.0. www.fws.gov/pacific/.../Albatross%20Action%20Plan%20 ver.1.0.pdf (accessed February 2008).

Nettleship, D. N., J. Burger, and M. Gochfeld, eds. 1994. *Seabirds on Islands: Threats, Case Studies and Action Plans.* BirdLife Conservation Series No. 1. Cambridge: BirdLife.

Nevard, Tim. 2005. "Big Bird Race, 2005." www.herbsphere.com/bbr1.htm (accessed May 2007).

Nevitt, Gabrielle, et al. 2004. "Testing Olfactory Foraging Strategies in an Antarctic Seabird Assemblage." *The Journal of Experimental Biology* 207: 3537–3544.

Nevitt, Gabrielle A., Marcel Losekoot, and Henri Weimerskirch. 2008. "Evidence for Olfactory Search in Wandering Albatross, *Diomdea exulans*." *Proceedings of the National Academy of Sciences of the United States of America* 105:12, 4576–4581.

New Zealand, Department of Conservation. 2007. "500th Royal Albatross Chick." Media release, February 8. http://www.doc.govt.nz/about-doc/news/media-releases/2007/500th-royal-albatross-chick-at-taiaroa-head-named/ (accessed July 2010).

———. 2009. "New Zealand's Migratory Species." http://www.doc.govt.nz/about-doc /role/international/migratory-species/ (accessed January 2010).

———. n.d. "Threats to Albatrosses." http://www.doc.govt.nz/templates/page.aspx ?id=32961 (accessed July 2010).

New Zealand, Ministry of Fisheries. 2008. "Seabird Protection Measures Announced." Feb. 21. http://www.fish.govt.nz/en-nz/Press/seabird+mitigation.htm (accessed March 2008).

New Zealand, Ministry of Fisheries and Department of Conservation. 2004. *National Plan of Action to Reduce the Incidental Catch of Seabirds in New Zealand Fisheries* (April). Wellington: DOC.

New Zealand, Ministry of Foreign Affairs. 2005. "Vessels Sighted on Antarctic Patrol Flight." Press release, December 16. www.colto.org/ArchiveNews.htm http:// www.google.com/search?ie=UTF-8&oe=UTF-8&sourceid=navclient&gfns=1&q =New+Zealand.++2005.+%E2%80%9CVessels+Sighted+on+Antarctic+Patrol+Fl ight.%E2%80%9D+Press+Release+16+December (accessed January 2008).

New Zealand, Oceans Policy Secretariat. 2003. "Environmental Issues." *Working Paper* 4, March (and other marine and wildlife issues). http://www.mfe.govt.nz/publica tions/oceans/nz-exclusive-economic-zone-discussion-paper-aug07/html/page5 .html (accessed May 2005 and onward).

New Zealand, Royal Albatross Center. 2010. Albatross Colony Reports. http://www .albatross.org.nz/news.html (accessed June 2010).

"New Zealand Recalls Squid Boats." 2005. May 6. http://en.wikinews.org/wiki/New_Zealand_recalls_squid_boats (accessed May 2006).

O'Byrne, David. 2010. "Early Withdrawal of Macquarie Island Pest Eradication Team." http://www.medoa.tas.gov.au/ (accessed July 2010).

Oceana. 2008. "The End of Pirate Fishing Vessel *Viarsa 1*." February 5. www.fishupdate.com/.../Oceana:_The_end_of_pirate_fishing_vessel_Viarsa_1.html (accessed January 2010).

Ocean Surface Currents in CIMAS. 2006. http://oceancurrents.rsmas.miami.edu (accessed January 2006).

OECD (Organization for Economic Cooperation and Development). "About the OECD." www.oecd.org.

The Organization for the Promotion of Responsible Tuna Fisheries. 2000. "Mission Statement." http://www.oprt.or.jp (accessed March 2008 and onward).

Ormerod, S. J. 2003. "Current Issues with Fish and Fisheries: Editor's Overview and Introduction." *Journal of Applied Ecology* 40: 204–213.

Orrego Vicuna, F. 1999. *The Changing International Law of High Seas Fisheries*. Cambridge: Cambridge University Press.

Owen, James. "Extinction Near for the Albatross, Experts Warn." *National Geographic News*, April 17.

Parkes, J. 2008. "A Feasibility Study for the Eradication of House Mice from Gough Island." Sandy: RSPB, *Research Report* No.34. http://www.rspb.org.uk/ourwork/conservation/projects/tristandacunha/publications.asp (accessed July 2010).

Parry, Gordon, and C. J. R. Robertson. 1998. *The Royals of Taiaroa*. Dunedin: Otago Peninsula Trust.

Pauly, Daniel. 1995. "Anecdotes and the Shifting Baseline Syndrome in Fisheries." *Trends in Ecology and Evolution* 10: 430.

Perrins, C. M., J. D. Lebreton, and G. J. M. Hirons, eds. 1991. *Bird Population Studies*. Oxford: Oxford University Press.

Peterson, M. J. 1993. "International Fisheries Management." In *Institutions of the Earth*, ed. Peter M. Haas et al., 249–305. Cambridge: MIT Press.

Pierre, Johanna. 2007. "Natural Deterrents." In Double 2007, 6.

Pinaud, David, and Henri Weimerskirch. 2002. "Ultimate and Proximate Factors Affecting the Breeding Performance of a Top-Predator." *Oikos* 99: 141–150.

Plumwood, Val. 2002. *Environmental Culture: The Ecological Crisis of Reason*. London: Routledge.

Pohl, Otto. 2003. "Challenge to Fishing: Keep the Wrong Species Out of Its Huge Nets." *New York Times*, July 29.

Pollard, R. T., M. I. Lucas, R. J. Sanders, and P. Statham. "Natural Iron Fertilization in the Southern Ocean: What Can It Tell Us About Carbon Export." www.soc.soton.ac.uk (accessed March 2007).

Poncet, Sally. 2006. "Report on Post-Eradication." www.southgeorgiasurveys.org/?download=Northeast%20Island%20post-rat%20eradication%20check%20September%202 (accessed March 2008).

Porcasi, Judith F. 1999. "Prehistoric Exploitation of Albatross on the Southern California Channel Islands." *Journal of California and Great Basin Anthropology* 21:1, 94–112. http://escholarship.org/uc/item/6hh657p5 (accessed July 2010).

Porter, Gareth, and Janet W. Brown. 1996. *Global Environmental Politics*. 2nd ed. Boulder: Westview Press.

Prickly Pear Control. http://australianscreen.com.au/titles/australasian-gazette-prickly/clip1/ (accessed December 2008).

Prince of Wales. 2004a. Address to Waterbirds Around the World Conference, Edinburgh, April 4–7. http://www.birdlife.org/news/news/2004/04/hrh_albatross .html (accessed June 2006).

———. 2004b. Letter to First Meeting of Parties, Hobart. November 10. http://www .birdlife.org/action/science/species/seabirds/clarence_house_letter.pdf (accessed November 2007).

———. 2005a. "Speech by HRH The Prince of Wales for the RSPB Albatross dinner, Trinity House, London." April 27. http://www.princeofwales.gov.uk/speech esandarticles/a_speech_by_hrh_the_prince_of_wales_for_the_rspb_albatross _d_55.html (accessed April 2009).

———. 2005b. "Speech Given at the Royal Albatross Center, Dunedin New Zealand." March 6. http://www.ecology.info/albatross.htm (accessed May 2006).

———. 2009. "HRH Hosts a Reception for The Albatross Task Force." http://www .princeofwales.gov.uk/newsandgallery/news/hrh_hosts_a_reception_for_the_al batross_task_force_at_claren_724987502.htm (accessed July 2010).

Radford, Tim. 2005. "Vital Marine Species Under Threat." *The Guardian*, July 29.

"Rat Eradication." 2002. *Wildlife Conservation in the Falkland Islands* (October 2): 8–9.

Raustiala, Kal. 1997. "States, NGOs, and International Environmental Institutions. " *International Studies Quarterly* 41: 719–740.

Rauzon, Mark K. 2001. *Isles of Refuge*. Honolulu: University of Hawai'i Press.

Rawls, John. 1971. *A Theory of Justice*. Cambridge: Belknap Press of Harvard University Press.

Revkin, Andrew C. 2003. "Commercial Fleets Slashed Stocks of Big Fish." *New York Times*, May 15.

Richdale, L. E. 1939. "A Royal Albatross Nesting on the Otago Peninsula, New Zealand." *The Emu* 28: 467–488.

———. 1942. "Supplementary Notes on the Royal Albatross." *The Emu* 41: 169–184.

———. 1950. "The Pre-egg Stage in the Albatross Family." *Biological Monographs* No. 3. Dunedin: Otago Daily Times.

———. 1952. "Post-Egg Period in Albatrosses." *Biological Monographs* No. 4. Dunedin: Otago Daily Times.

Rickard, L. S. 1965. *The Whaling Trade in New Zealand*. Auckland: Minerva.

Ridgway, John. 2003–2004a. "Daily Log." http://www.bbc.co.uk/dna/h2g2/brunel /A894594.

———. 2003–2004b. "John Ridgway Save the Albatross Voyage." http://www.bbc.co .uk/dna/h2g2/brunel/A894594 (accessed May 2006).

Robertson, C. J. R. 1991. "Questions on the Harvesting of Toroa on the Chatham Islands." *Science & Research Series* 35. Wellington: Department of Conservation.

———. 1993a. "Effects of Nature Tourism on Marine Wildlife." *Proceedings of a Conference on Marine Conservation and Wildlife Protection,* 53–60. Wellington: New Zealand Conservation Authority.

———. 1993b. "Survival and Longevity of the Northern Royal Albatross *Diomedea epomophora sanfordi* at Taiaroa Head." *The Emu* 93: 269–276.

———. 1998a. "Factors Influencing the Breeding Performance of the Northern Royal Albatross." In Robertson and Gales 1998a, 99–104.

———. 1998b. "Richdale, Lancelot Eric 1900–1983." In *The Dictionary of New Zealand Biography,* Vol. 4, 430. Auckland: Auckland University Press.

———. 2001. "Effects of Intervention on the Royal Albatross Population at Taiaroa Head, Otago, 1937–2001." New Zealand Department of Conservation (DOC). Science Internal Series 23. Wellington: DOC.

Robertson, C. J. R., E. A. Bell, N. Sinclair, and B. D. Bell. 2003. *Distribution of Seabirds from New Zealand That Overlap with Fisheries Worldwide.* Science for Conservation 233. Wellington: Department of Conservation.

Robertson, C. J. R., and G. B. Nunn 1998. "Towards a New Taxonomy for Albatrosses." In Robertson and Gales 1998a, 13–19.

Robertson, C. J. R., and A. Wright. 1973. "Successful Hand-Rearing of an Abandoned Royal Albatross Chick." *Notornis* 20: 49–58.

Robertson, Graham. 1998. "The Culture and Practice of Longline Tuna Fishing." *Bird Conservation International* 8: 211–221.

———. 2002. "Can Albatrosses and Long Line Fisheries Co-Exist?" *Australian Antarctic Magazine,* December 2.

———. 2006. " Fast Sinking (Integrated Weight) Longlines." *Biological Conservation* 132: 458–471.

———. 2007. "Underwater Bait-Setting." Report of First Meeting of the Seabird Bycatch Working Group, Third Meeting of Advisory Group (To ACAP), Valdivia, Chile, June 19–22, p. 5. http://www.acap.aq/en/index.php?option=com_docman&task=cat_view&gid=66&Itemid=33 (accessed March 2008).

Robertson, Graham, and Rosemary Gales, eds. 1998a. *Albatross: Biology and Conservation.* Chipping Norton, NSW: Surrey Beatty and Sons.

———. 1998b. "Preface." In Robertson and Gales 1998a, vii–ix.

Robertson, Graham, Malcolm McNeill, Neville Smith, Barbara Wienecke, Steven Candy, and Frederique Olivier. 2006. "Fast Sinking (Integrated Weight) Longlines Reduce Mortality of White-chinned Petrels (*Procellaria aequinoctialis*) and Sooty Shearwaters (*Puffinus griseus*) in Demersal Longline Fisheries." *Biological Conservation* 132: 458–471.

Roheim, Cathy A. 2003. "The Seafood Consumer." In Anderson 2003, 194–204.

RSPB (Royal Society for the Protection of Birds). 2000. "British Bird Watching Fair." *Birds* (Autumn): 35–36.

———. 2008. "Sir David Lends His Support." RSPB Tweet. http://www.rspb.org.uk /tweet/winter08/sirdavid.asp (accessed December 2009).

Safina, Carl. 2002. *Eye of the Albatross*. New York: Macrae.

Salkeld, Luke. 2007. "Look What's Blown In." *Daily Mail* (UK), July 3.

Sample, Ian. 2005. "Census Reveals Oceans' Secrets." *The Guardian*, December 15.

———. 2006. "Deep Sea Fish Face Extinction." *The Guardian*, January 5.

Sanchirico, James N. 2004. "A Social Scientist's Perspective on the Potential Benefits of the Census of Marine Life." *RFF Discussion Paper* 04–23 rev., June.

"Save the Albatross." 2002. *Birds* 19: 3 (Autumn): 34–40.

Seafood Industry Council. 2006. "Southern Seabird Solutions Fisherman Off to Peru." http://www.seafoodindustry.co.nz/n379,44.html (accessed October 2008).

Serventy, D. L., V. N. Serventy, and J. Warham. 1971. *The Handbook of Australian Sea-Birds*. Sydney: Reed.

Short-tailed Albatross. Press releases. http://www.birdlife.org/news/news/2008/03 /start_translocation.html (accessed December 2008).

Sill, Edward R. 1944. *Around the Horn*. New Haven: Yale University Press.

Small, Cleo J. 2005. *Regional Fisheries Management Organisations: Their Duties and Performance in Reducing Bycatch of Albatrosses and Other Species*. Cambridge: BirdLife International.

Smith, R. I. Lewis. 1995. "Conservation Status of South Georgia and the Sandwich Islands." In Dingwall 1995, 3–14.

Smith, T. 2000. "Seabird Bycatch Reduction in Long Line Fisheries: An Industry Perspective." *Marine Ornithology* 28: 2.

South Africa, Department of Environmental Affairs and Tourism. 2008. *National Plan of Action for Reducing the Incidental Catch of Seabirds in Longline Fisheries*. NPOA-pdf.

South Africa, State of the Environment. 2005. "The Prince Edward Islands." http:// soer.deat.gov.za/themes.aspx?m=252 (accessed July 2008).

South Georgia Heritage Trust. 2008. http//www.sght.org (accessed June 2007).

Southern Seabird Solutions. 2002."Ground-breaking Seabird Alliance." Media release, July 19, doc.govt.nz/Conservation. http://www.southernseabirds.org/n1043, 183.html (accessed August 2004).

———. 2003. "Seabird Device to Be Tested." Media release, May 26. http://www.south ernseabirds.org/n1043,183.html (accessed July 2010).

———. 2004. "Slick Experiment Moving Along." *Newsletter*, December, p. 3. Ø97– southern-seabird-solutions-newsletter-dec-2004.pdf.

———. 2005. "Reunion Island Update." *Newsletter*, December 3 (and "Looking Ahead." *Newsletter*, June 2004, p. 6) http://www.southernseabirds.org/ss-news_and_events (accessed November 2007).

———. 2006. "Fisherman Dave Kellian Visits Peru." *Newsletter*, October 6. http://www .southernseabirds.org/ss-news_and_events (accessed November 2007).

———. 2007. "500th Royal Albatross Chick." Media release, February 8.

———. 2008. " Southern Seabird Solutions Trust." http://www.southernseabirds.org/ (accessed March 2008).

——. 2010. "About Us." http://www.southernseabirds.org/ss-about_us (accessed July 2010).

Sovacool, Benjamin K., and Kelly E. Siman-Sovacool. 2008. "Creating Legal Teeth for Toothfish: Using the Market to Protect Fish Stocks in Antarctica." *Journal of Environmental Law* 20: 15–33.

Steiner, Rick. 1998. "Resurrection in the Wind: Comeback of the Short-tailed Albatross." *International Wildlife* (September–October): 4.

Stevens, William K. 1996a. "The Effectiveness of CCAMLR." *Governing the Antarctic.* In Stokke and Vidas 1996, 120–151.

——. 1996b. "Long-Line Fishing Seen as Damaging to Some Fish and to the Albatross." *New York Times*, November 5.

Stokke, Olav S. 2001a. "Conclusions." In Stokke 2001b, 329–360.

——, ed. 2001b. *Governing High Seas Fisheries.* Oxford: Oxford University Press.

Stokke, Olav S., and Davor Vidas, eds. 1996. *Governing the Antarctic.* Cambridge: Cambridge University Press.

Strange, Ian J. 1983. *The Falkland Islands.* 3rd ed. Newton Abbot: David and Charles.

Sullivan, B. J., T. A. Reid, and L. Bugoni. 2006. "Seabird Mortality on Factory Trawlers in the Falkland Islands." *Biological Conservation* 131:495–504.

TAAF (Terres Australes et Antarctiques Françaises). 2007. TAAF Presentation. www.taaf.fr/spip/spip.php?article177 (accessed February 2008).

TenBruggencate, Jan. 2005. "New Clean up Efforts Fight Ocean Debris." honoluluadvertiser.com/article/2005/Oct/12/ln/FP510120334.html (accessed January 2007).

Terence, Lindsay. 2008. *Albatrosses.* Australian Natural History Series. Collingwood: CSIRO Publishing.

TerraNature Trust. 2003. "Government Plan to Save Endangered Seabirds Is Not Good Enough." August 14. http://www.terranature.org/seabirds.htm (accessed November 2006).

Thorpe, William H. 1974. "David Lack, 1910–1973." *Biographical Memoirs of Fellows of the Royal Society* 20 (December): 271–293.

Tickell, W. L. N. 1970. "The Great Albatrosses." *Scientific American* 223 (November): 84–89.

——. 1975. "Observations on the Status of Steller's Albatross (*Diomedea albatrus*) 1973." *Bull ICBP* 12: 125–131.

——. 2000. *Albatrosses.* New Haven: Yale University Press.

Tickell, W. L. N., and J. D. Gibson. 1968. "Movements of Wandering Albatrosses (*Diomedea exulans*)." *The Emu* 68 (June): 7–20.

Tickell, W. L. N., and Philip Morton. 1977. "The Albatross of Torishima." *Geographical Magazine* 49: 359–363.

Tinbergen, Nikolaas. 1953. *The Herring Gull's World.* London: Collins.

Tolba, Mostafa K., and Iwona Rummel-Bulska. 1998. *Global Environmental Diplomacy.* Cambridge: MIT Press.

Traffic. 2008. "Illegal Toothfish Still on the Plate." http://www.traffic.org/home/2008/11/5/illegal-toothfish-still-on-the-plate.html (accessed November 2009).

Tuck, Geoffrey N., Tom Polachek, and Cathy M. Bulman. 2003. "Spatio-temporal Trends of Longline Fishing Effort in the Southern Ocean and Implications for Seabird Bycatch." *Biological Conservation* 114:1 (November): 1–27.

Tuck, Geoffrey N., Tom Polachek, J. P. Croxall, and Henri Weimerskirch. 2001. "Modelling the Impact of Fishery By-Catches on Albatross Populations." *Journal of Applied Ecology* 38: 1182–1196.

UNCLOS (United Nations Convention on the Law of the Sea). n.d. List of Parties. http://www.un.org/Depts/los/reference_files/chronological_lists_of_ratifications.htm (accessed July 2010).

Underdal, Arild, ed. 1998. *The Politics of International Environmental Management*. Dordrecht: Kluwer.

United Nations. 1982. "Comment on the Law of the Sea, Part VII, High Seas." http://www.un.org/Depts/los/convention-arguments/texts/unclos/ (accessed July 2010).

United Nations, Environmental Program. 2002. "The Prince of Wales Urges Governments to Back International Albatross Conservation Deal." Nairobi, Kenya, September 18. athttp://www.unep.org/Documents.Multilingual/Default.asp?DocumentID=264&ArticleID=3132&l=en (accessed September 2007).

UN, FAO. 1993. The Agreement to Promote Compliance with International Conservation and Management Measures by Fishing Vessels on the High Seas ("Compliance Agreement"), International Fisheries Agreements online www.oceanlaw.net/projects/current/pdf/ifa_sample.pdf (accessed July 2010).

——. 1995. "Code of Conduct for Responsible Fisheries." http://www.fao.org/DOCREP/005/v9878e/v9878e00.htm (accessed March 2007).

——. 2000–08. Fisheries and Aquaculture Department, Regional Fisheries Bodies. http://www.fao.org/fishery/rfb (accessed July 2010).

——. 2003a. Country Profiles: New Zealand. http://www.fao.org/fi/fcp/en/NZL/profile.htm and Fishery Sector (accessed March 2007).

——. 2003b. "Statistics. New Zealand Products." *Physical Flow Account for Fish Resources in NZ, 1998–2001*. Rome: FAO.

——. 2004. *Report of the Technical Consultation to Review International Plan of Action to Prevent Illegal, Unreported and Unregulated Fishing*, etc. Fisheries Report 753. Rome: FAO. http://www.fao.org/docrep/007/y5681e/y5681e00.HTM (accessed April 2007).

——. 2005. *Review of the State of World Marine Fishery Resources*. Fisheries Technical Paper 457. Rome: FAO.

——. 2008. "Report of the Expert Consultation on Best Practice Technical Guidelines," Fisheries and Aquaculture Report No. 880. Rome: FAO.

——. 2010a. Fisheries and Aquaculture Department, "Implementation of the 1995 FAO Code of Conduct." http://www.fao.org/fishery/ccrf/2/en (accessed July 2010).

——. 2010b. Regional Fisheries Bodies, Fact Sheets. http://www.fao.org/fishery/rfb/search/en (accessed July 2010).

UN, FAO, Committee on Fisheries. 2009. "Progress in the Implementation of the Code of Conduct." COFI/2009/2. Rome: FAO.

UN, FAO, Fisheries and Aquaculture Department. 2005. *Incidental Catch of Seabirds in Longline Fisheries.* (Text by John W. Valdemarsen). Rome: FAO http://www.fao.org /fishery/topic/14803/en (accessed December 2008).

US Fish and Wildlife Service. 2005. *Short-tailed Albatross Draft Recovery Plan.* Anchorage, AK: Government Printing Office.

Upton, Simon, and Vangelis Vitalis. 2003. "Stopping the High Seas Robbers: Coming to Grips with Illegal, Unreported and Unregulated Fisheries on the High Seas." Round Table on Sustainable Development. Paris: OECD. http://www.illegal-fish ing.info/uploads/OECDpaper-roundtable-on-SD.pdf. (accessed June 2007).

Viarsa 1. 2003. News.com. http://www.colto.org/Articles_Latest_Viarsa.htm (accessed June 2007).

Vidal, John. 2006. "When the Boat Comes In." *The Guardian*, November 6.

———. 2008. "From Stowaway to Supersize Predator." *The Guardian*, May 20 (quote). http://www.scar.org/ (accessed August 2010).

Vogler, John. 1995. *The Global Commons.* Chichester, UK: Wiley.

"Voluntary Action Fails to Save Seabirds." 2005. BBC News, May 6.

Wainwright, Martin. 2006. "Artists Champion Fishermen's Cause." *The Guardian*, January 2.

Walker, Matt. 2009. "Albatrosses Set Breeding Record." BBC, Earth News, Sept. 2. http:// news.bbc.co.uk/earth/hi/earth_news/newsid_8232000/8232337.stm (accessed August 2010).

Wallace, Ian. 2004. *Beguiled by Birds.* London: Helm.

Walton, D. W. H., and W. N. Bonner. 1985. "History and Exploration in Antarctic Biology." In Bonner and Walton 1985, 16.

Ward, Veronica. 1998. "Sovereignty and Ecosystem Management." In *The Greening of Sovereignty in World Politics*, ed. Karen T. Litfin, 79–108. Cambridge: MIT Press.

Warham, John. 1990. *The Petrels: Their Ecology and Breeding Systems.* London: Academic Press.

———. 1996. *The Behaviour, Population Biology and Physiology of the Petrels.* London: Academic Press.

Waring, George H., and Chandler S. Robbins. 1996. "In Memoriam: Harvey I. Fisher, 1916–1994." *The Auk* 113: 928–930.

"The Waterbird Society." n.d. http://www.waterbirds.org/murphy.html (accessed February 2007).

Watson, Reginald, and Daniel Pauly. 2001. "Systematic Distortions in World Fisheries Catch Trends." *Nature* 414 (November 29): 534–536.

WCPFC (Western and Central Pacific Fisheries Commission). 2006a. "Decisions of the Commission." http://www.wcpfc.int/wcpfc4/index.htm (accessed June 2007).

———. 2006b. "Report and Information Paper of the Voluntary Small Working Group on Seabird Bycatch Mitigation." http://www.wcpfc.int/wcpfc4/index.htm (accessed 2007).

———. 2007. "Report on the Work of the Commission." http://www.wcpfc.int/wcpfc4 /index.htm (accessed March 2008).

Weimerskirch, Henri. 2004. "Wherever the Winds May Blow." *Natural History*, October, 40.

Weimerskirch, Henri, and Pierre Jouventin. 1987. "Population Dynamics of the Wandering Albatross (*Diomedea exulans*) of the Crozet Islands." *Oikos* 49: 315–322.

Wigan, Michael. 1998. *The Last of the Hunter Gatherers*. Shrewsbury: Swan Hill.

Williams, Stanley Thomas, and Barbara D. Simison. 1944. *Around the Horn*. New Haven: Yale University Press.

Willock, Anna. 2002. *Uncharted Waters*. Cambridge, UK: Traffic International.

Winegrad, G., and G. E. Wallace. 2005. *Stopping Seabird Bycatch*. Washington, DC: American Bird Conservancy. http://www.abcbirds.org/index.html (accessed March 2008).

Witherby, H. F., F. C. R. Jourdain, N. F. Ticehurst, and B. W. Tcker. 1940. *The Handbook of British Birds*. 5 vols. London: Witherby.

Wolfrum, Rudiger, V. Roben, and Fred L. Morrison. 2000. "Pressure on the Marine Environment." In *International Regional and National Environmental Law*, ed. Fred Morrison and R. Wolfrum, 225–283. The Hague: Kluver Law.

Woods, Robin W., and Anne Woods. 1997. *Atlas of Breeding Birds of the Falkland Islands*. Oswestry: Nelson.

World Heritage. n.d. whc.unesco.org/en/about/. Also see WH List (accessed February 2008 and onward).

World Wildlife Fund. 2009. "The Underwater Baited Hook." 2009 Grand Prize Winner ($30,000). http://www.smartgear.org/smartgear_winners/2009/grand_prize_winner_2009/ (accessed July 2010).

Worm, Boris, et al. 2006. "Impacts of Biodiversity Loss on Ocean Ecosystem Services." *Science* 314:5800 (November 3): 787–790.

Worm, Boris, Ray Hilborn, et al. 2009. "Rebuilding Global Fisheries." *Science* (July 31): 578–585.

Young, Oran R. 1989. *International Cooperation: Building Regimes for Natural Resources and the Environment*. Ithaca: Cornell University Press.

———. 1994. *International Governance: Protecting the Environment in a Stateless Society*. Ithaca: Cornell University Press.

———. 1998. *Creating Regimes*. Ithaca: Cornell University Press.

———. 2002. *The Institutional Dimensions of Environmental Change*. Cambridge: MIT Press.

Young, Oran R., George J. Demko, and Kilaparti Ramakrishna, eds. 1996. *Global Environmental Change and International Governance*. Hanover: University Press of New England.

Young, Oran R., and Gail Osherenko, eds. 1993. *Polar Politics*. Ithaca: Cornell University Press.

Index

Page numbers in *italics* indicate photos.

Australia, 13–15, 28, 56, 92, 94, 112, 119–124, 199–206, 208–209, 212, 220, 229, 231, 235, 237; albatross habitats in, 53, 59, 67–70, 90, 129, 214; commercial fishing in, xx, xxiii, 113, 122–124, 133, 135, 156–158, 160–162, 194–195, 216; conservation efforts in, xviii–xix, xxi–xxii, 85, 87, 97–98, 100, 123, 142, 151–152, 154, 160, 179, 188, 192, 200, 218, 240–243, 245, 249; study of albatrosses in, xvii, xix, 45–50, 60, 77, 89, 132, 136–138, 164–166

Australian Antarctic Division, 138, 164–165, 200

Australian Marine Conservation Society, 220

Aves Uruguay, 223

Baker, Barry, 166, 205
Banks, Joseph, 13, 21
Barstow, Grace Emeline, 22, 245
Battam, Henry, 46, 50
Bayly, William, 14
Bellambi (Australia), 46–47, 49
Bellamy, David, 233
Benson-Pope, David, 212
Berguno, Jorge, 189–190
Berkner, Lloyd, 93
Big Bird Race, 233–234
Bird Island (UK), 27, 45–50, 48, 58, 70, 132, 136, 238
BirdLife International, xix, xxii–xxiv, 4, 155, 160, 163, 180, 221, 234–235, 266n2; and ACAP, 152, 241; Albatross Task Force, xxiv, 222–223, 225, 236; and fishery management organizations, 194–197, 224; and Forest & Bird, 207, 237; Global Seabird Program, xxii–xxiii, 132, 178, 180, 221, 223; Important Bird Areas (IBA) Program, 221–222; and the IOTC, 195; and IPOA—Seabirds, 179, 224; Marine Programme, 216–218; and NGOs, 220–221; and population of albatrosses, 70, 75, 137, 140, 144, 225; Save the Albatross campaign, xxiii–xxiv, 221, 233, 235–236;

and taxonomy of albatrosses, 63, 262n1; *Tracking Ocean Wanderers*, xxiv, 128, 196, 221; and the WCPFC, 193

Black-browed Albatross, 14, 26, 58, 64, 69, 73–76, 74, 98, 135, 161, 201, 213, 214–216, 223

Black-footed Albatross, 2, 17, 19, 36, 43, 64, 72, 214

blast fishing, 141–142

Blue Ocean Institute, 38

Bluefin Tuna, xviii, xx, 51, 92, 100, 104, 106, 112–115, 123–125, 133, 136, 140, 158, 165, 193–194, 201, 217, 242–243, 261n10

Bonin Islands, 19, 32, 42, 43, 67, 72, 97

Bonn Convention on the Conservation of Migratory Species of Wild Animals (CMS) (1979), xviii–xix, xxi, xxiii, 85–88, 137, 141, 152, 165, 168, 203–205, 235, 240–242

Bounty Island (NZ), 57, 69

Bours, Helen, 230

Brazil, 165, 190, 196, 214, 218, 221, 223–224, 226, 241, 244–245

British Antarctic Survey, 25, 48, 50, 58, 62, 89, 127, 130, 132

British Ornithologists' Union, 82, 130–131

British Overseas Territories Act (2002), 57–58

British Trust for Ornithology, 29

Brothers, Nigel, xix, xxi–xxii, 127, 132–138, 140, 156–158, 160–162, 163–166, 178, 190, 194, 205

Brown, Derek, 215

Brown Rat, 26, 209

Buller's Albatross, 33, 68, 70, 214, 257n1

Burgmans, Anthony, 172–173

Burhenne-Guilmin, Francoise, 86

"Button," 44, 238, 238, 259n1

bycatch, xiv–xv, xix–xxiv, 3, 92–93, 106–107, 114, 118, 122–123, 126, 133, 135, 138–139, 140–145, 153–164, 174–182, 185, 188–197, 200, 207–208, 210–211, 213, 215–218, 220–232, 236, 240, 245, 247–251; limits, 205, 211, 223

THE ALBATROSS AND THE FISH

229, 244; International Plan of Action—
Illegal, Unreported and Unregulated
Fishing, xxiii, 172, 179; International Plan
of Action—Seabirds, xxii, 169, 171–172,
178–180, 192, 195, 224, 240, 242. *See also*
United Nations Code of Conduct for
Responsible Fisheries
United Nations General Assembly, xix, 82,
86, 142
United States, 18–19, 35, 72, 82–85, 87, 93, 95,
100, 102, 106, 113, 143, 146, 151, 165–167,
175–177, 180, 188, 199, 204, 212, 218, 220,
231, 233, 241–242, 245
United States Fish and Wildlife Service, 83
United States Marine Mammal Protection
Act (1972), 141
United States Western Pacific Regional
Fishery Management Council, 231
University of Oxford. *See* Edward Grey In-
stitute of Field Ornithology
Uruguay, xxiii, 165, 186–187, 203, 218, 223–
225, 232, 241
USSR, 85, 87, 94, 111, 119, 136, 207

Viarsa 1, xxiv, 186, 203–204, 228

Wandering Albatross, xxi, 14–15, 23–25,
27–28, 33, 44–50, 57, 62–67, 66, 70, 98,
128–130, 131, 136, 201, 204, 213, 238
Watson, Reginald, 104
Waved Albatross, 35, 51, 64, 67–68, 71, 99,
137, 200, 214, 222
Weimerskirch, Henri, 50, 127–129, 138, 165,
196
Western and Central Pacific Fisheries Commission (WCPFC), 118, 123–124, 191–193,
195, 231, 249

whaling, 14–16, 25–28, 50, 71, 85, 87, 95, 111,
127, 129, 151, 182, 208, 247
White-capped Albatross, 69–70, 144, 244
Wildlife Act (1953, NZ), 31
wildlife refuges, 31, 38, 72, 83, 199
Willock, Anna, 100
World Albatross Conference (1995), xx,
xxii–xxiv, 162, 163–165, 204
World Conservation Union. *See* Interna-
tional Union for the Conservation of
Nature
World Fund for Nature, 196, 221
World Heritage Convention, xviii, 95–99,
241. *See also* International Union for the
Conservation of Nature (IUCN); World
Heritage Sites
World Heritage Sites, xviii, 43, 54, 71, 96–99,
146, 153, 200–201, 209, 222
World War I, 16, 18, 30, 102
World War II, 16, 18, 29, 31, 33, 88–89, 91, 93,
103, 113–114, 199, 229
World Wildlife Fund (WWF), 115, 172, 176,
191, 229, 231
Worm, Boris, 89, 104, 125, 198, 261n7
Wright, Alan, 31

Yamashina Institute for Ornithology, 40
Yellow-nosed Albatross, 64, 68, 73, 75, 213,
214, 216

Zoological Society: of London, 28, 131; of
New York, 40